OTHER BOOKS BY THIS AUTHOR

ENERGY AND GLOBAL WARMING -
How to ensure human survival

Dieter H. Otterbach, Ed. Patria, 2013

A WORLD ON THE MARCH - Universal History Review

Dieter H. Otterbach and Irma López Martinez, 2016

GREAT CHARACTERS - Testimony of Fascinating Lives, Vol I.

Irma López Martinez and Dieter H. Otterbach, 6th edition, 2020

GREAT CHARACTERS - Testimony of Fascinating Lives, Vol. II.

Irma López Martinez and Dieter H. Otterbach, 2021

CLIMATE CHANGE
AND
CLEAN ENERGY

DIETER H. OTTERBACH

authorHOUSE®

AuthorHouse™
1663 Liberty Drive
Bloomington, IN 47403
www.authorhouse.com
Phone: 833-262-8899

Published by AuthorHouse 08/09/2021

ISBN: 978-1-6655-3441-3 (sc)
ISBN: 978-1-6655-3442-0 (hc)
ISBN: 978-1-6655-3449-9 (e)

Library of Congress Control Number: 2021916188

To contact the author write to:
dieterho@prodigy.net.mx

CONTENTS

ENERGY AND CLIMATE CHANGE

OBJECTIVE

The rapid progress of global warming presents a serious and imminent threat to humanity. The purpose of this book is to indicate the many causes and effects of climate change. To define potential actions to overcome this danger, we have to examine all possible forms of energy available on our planet, because this force alone controls all life on Earth. Reducing the use of fossil fuels to the absolute minimum possible is essential to reducing greenhouse gas emissions to the level that the existing biosphere can absorb. The most in-depth analysis of all existing and developing renewable clean energies concluded that the exponential growth of the human population causes a 45% deficit of clean electricity, which can only be satisfied through processes that still produce greenhouse gases. Only two nuclear processes, nuclear fission and fusion, in no way contribute to climate change. The fission of uranium, U-235, slightly enriched, has shown very significant advances with some prototypes. However, several disadvantages indicate that is not a long term solution.

Today's great hope is nuclear fusion, the same process the Sun uses to generate its enormous energy. A lot of research has already made significant progress toward turning this reaction into a commercial process that promises clean, safe and abundant electrical energy to free the world from the threat of global warming.

In addition, three wonderful natural processes, photosynthesis, the conversion of seawater into drinking water, and the production of natural fertilizer deserve in-depth investigation because of their promise to eliminate the inevitable emission of greenhouse gases associated with industrial processes.

ENERGY AND GLOBAL WARMING

INTRODUCTION

More and stronger hurricanes; catastrophic fires throughout the western United States, with the loss of dozens of lives and more than a thousand homes turned to ash; caravans of immigrants fleeing droughts in Central America, each with up to a thousand refugees migrating through Mexico to the North; island republics, such as Kiribati, Malvinas and others, considering to evacuate villages: These are just a few examples of how global warming is manifesting itself today. If we are not able to limit the increase in global temperature to a maximum of 1.5^0C (centigrade) within twenty years, the warming will be irreversible, with consequences that put the survival of humanity at risk. NASA has shown that the melting of the thick ice caps at both poles is accelerating, which will result in a rise in sea level of 1.5m in less than 20 years. Millions of climate refugees from large coastal cities (New York, Miami, Shanghai, Rio de Janeiro, etc.) will have to move to higher ground. Glaciers all over the world are in the process of disappearing or no longer exist. Most of the great rivers of the world have their origin in high-mountain glaciers; in Europe, in certain seasons and regions, the Rhine and Danube rivers are already at such low levels that navigating them is impossible. On the other side of the world, the great rivers that originate in the Himalayas - the Indus, Ganges, Brahmaputra, Mekong and Yangtze - have seen higher flows in recent years. These will be reduced to a minimum with the thaw in these mountain ranges, with unimaginable effects for the billions of inhabitants of India, Bangladesh and the countries of Indochina.

Global warming will become increasingly noticeable in the following areas as well: agriculture and forestry, biodiversity, lack of water, more and new pests, higher food costs and, ultimately, danger to our health.

By now, hardly anyone doubts the imminent danger these threats represent to us or, at least, to our descendants. A few politicians continue to deny this reality, mainly so as not to put their position or reelection at risk. Others propose drastic or absurd laws without consulting with scientists. For example, eliminating all fossil fuels within fifteen years would leave us completely without many products absolutely essential for life today, such as steel, cement, plastics, synthetic fibers, fertilizers, etc., in turn leading to a lack of food, as will be explained further on in this text.

The fundamental cause of the excessive warming of our Earth is the burning of fossil fuels, mainly in electricity production and transportation, generating carbon dioxide (CO_2) and causing a progressive greenhouse effect. It is generally assumed that if we generated all electrical energy using natural, clean and renewable resources, we could stop global warming, because all transportation would be converted to use electrical energy. With enough clean and renewable electrical energy, we could drastically reduce the generation of greenhouse gases. We could even produce potable water from the sea to meet the needs of human life. Unfortunately, as this book shows, this concept is not realistic in the absence of profound change and expensive technical developments.

This book will clarify that the growth of the world population has produced such a demand for electrical energy that all known natural resources could never satisfy it. For this reason, we will discuss here an idea that dates back to the 1920s, when it was promised to become a reality within thirty years: nuclear fusion, the equivalent of creating a small sun here on Earth to produce clean and abundant electrical energy without greenhouse gases or radioactive products. In the following chapters, we list three well-known natural processes that convert salt water into fresh water, generate fertilizer from air, and produce photosynthesis, all under ambient conditions. These miraculous natural processes deserve in-depth scientific research to replace highly polluting industries with natural processes that are much more benign for our environment.

I

WORLD ENERGY SOURCES AND USES

To begin, we must first learn about the full range of energy sources available on the planet, how they are used, and even how they are abused if we are to understand their economic impact and potential for contamination. These sources are classified into a) energy that comes directly from the Sun, such as solar radiation and its use in solar panels and wind turbines; b) energy that the Sun produced here on Earth, millions of years ago, in the form of fossil fuels; c) energy from the Earth, such as geothermal, nuclear and gravitational energy; and d) renewable and non-renewable energy.

1) FOSSIL FUELS, NON-RENEWABLE ENERGY FROM THE SUN

Coal, oil, and natural gas are the products of the anaerobic decomposition of ancient plant and animal life on Earth. Photosynthesis, which is the interaction of sunlight, water and CO_2 in plants, gave rise to this material. In subsequent geological processes during the Carboniferous period (340 million years ago), these remains were converted into hydrocarbons, chemical compounds that contain only hydrogen and carbon atoms. Today, fossil fuels, which are a form of condensed energy from the Sun, are still the main source of electrical energy and power for almost all transportation. The burning of these fuels produces the greenhouse gas, CO_2, responsible for global warming and for polluting the atmosphere. These energy resources are not renewable, but new extraction techniques (fracking) ensure, at least, more than half a century

1

of availability, until new and clean renewable energy sources can reduce their use to a minimum. This book provides a detailed overview of the generation and use of this energy.

2) ENERGY FROM THE EARTH

When classifying the various types of this energy, both geothermal and nuclear forms are considered non-renewable, because geothermal wells are gradually depleting and nuclear energy depends on the extraction of uranium or other radioactive minerals. However, these two carbon-free energy sources promise to have a very long life, because many sites in volcanic areas have not yet been exploited for their energy potential. Uranium and other radioactive minerals (thorium) appear to be abundant, and nuclear breeder reactors produce more fissile material than they consume. More detail on using these two important resources is provided in the following chapters.

– Gravitation of the Earth

Gravity is the attraction between masses and, as such, it is a form of energy. It is a force always present in the Universe and on Earth. This invisible force keeps the Earth and the planets in their orbits around the Sun. It causes everything to fall towards the center of the Earth and keeps our feet on the ground. Gravity controls the flow of rivers downstream. Its energy causes the water from the top of dams to flow through turbines at the bottom, generating electricity. In fact, it takes a lot of energy to overcome Earth's gravity; for example, to launch a satellite into space via rockets.

– Earth's magnetic field

Earth's magnetic field is caused by electrical currents that form in its rotating inner core of iron and nickel. This magnetosphere extends for several thousand kilometers above our planet into space. A magnetic field can be considered a source of energy, but on Earth it is too weak to be considered a viable source of energy. Nevertheless, this magnetic shield protects the Earth from harmful cosmic radiation, deflecting it towards the poles, where it manifests as the aurora borealis in the north and aurora australis in the south. On Earth and in space, the magnetic field is used for navigation (compass, etc.) and for altitude control for orbiting satellites. Even migrating birds are oriented by this magnetic field.

Still, a small flow of useful electrical energy can be generated in space: When a wire conductor passes through a magnetic field, electrons flow in that wire. That's how electric generators work. A copper coil has been shown to generate enough electricity to recharge a battery when a spacecraft circles through the planet's magnetic field, and this technology is used in some satellites.

3) RENEWABLE ENERGY

Renewable energy is humanity's great hope. It is generated directly or indirectly from the Sun. The Sun's light is used directly to heat water flowing through panels or focused with concave mirrors to create very high heat. All energy related to the movement of water, such as tidal energy or wave energy, is gravitational energy from the Moon and the Sun. Hydropower may seem eternal, but climate change, geological events, and human activity, such as deforestation, could in time limit this energy. Using sunlight to generate electricity directly through solar panels, called photovoltaics, is possible only during the day. Ensuring a constant supply requires high-capacity and long-life batteries, an area that still requires new technology.

A general concern about renewable energy sources is their variable and intermittent availability, particularly solar and wind (day/night). These problems require smart solutions, such as integration into a large power grid and digitally controlled power flows. Their great advantage, of course, is the long-term sustainability of Sun-based renewable energy and the absence of greenhouse gas emissions. When available renewable energy cannot meet demand, gas-fired thermoelectric plants can fill in for them in the short term.

4) ENERGY FROM THE UNIVERSE

Cosmic rays have the highest energy, but they still have no practical use. This radiation constantly impacts the Earth. It consists of high-energy atomic particles, such as protons, helium nuclei, neutrinos, and various other forms of radiation that are emitted from the Sun's energy processes, originating primarily from sunspots that have an eleven-year cycle. But many more particles come from the active stars in our galaxy, the Milky Way, as well as from the farthest reaches of the Universe, from neutron stars, black holes and, especially, supernovae. This radiation has energy greater than 10^{20}eV (electron volts), much higher than particles made in the largest Earth based accelerator.

The observation and, especially, the forecasting of these high-energy flows in the Universe are carried out at the Pierre Auger Observatory in Argentina. Some of these particles become secondary when they interact with the Earth's atmosphere and its magnetic field, which provides a shield for humans, animals and vegetation on our planet. In space, without this protection, astronauts have seven minutes to seek shelter in a lead-lined cell when a solar storm is detected. These solar flares are known to be responsible for critical broad spectrum power failures, transformer explosions, and distortions in many communication systems. Supernova explosions in our galaxy, and in others nearby, can be dangerous for Earth as this energy can cause various types of cancer and is highly erratic and variable.

II

WORLD ENERGY - PRODUCTION AND DEMAND

Energy, water and air created life on Earth. Without energy, nothing could move or grow. Our planet would be dead. The most extensive and inexhaustible source of energy on Earth is radiation from the Sun in the form of light (short waves), and heat (infrared waves), which formed fossil fuels. The use of these resources must be limited because they produce greenhouse gases. To curb climate change, it is imperative that we reduce the use of coal and hydrocarbons to the essential minimum, converting all land transport to electricity. Thus we are faced with the great challenge of living mainly from the Sun's radiation and converting it into clean, renewable electrical energy. Just 0.2% of all energy that comes from the Sun to the Earth could satisfy global energy demand. The goal is to use appropriate and affordable technology to make the most of this resource and generate electricity to meet all global demand.

To understand the enormity of this goal, we must first understand the annual use and consumption (2019) of primary energy and how it is obtained:

- Coal: 25.2%, mostly for thermoelectric plants
- Oil: 34.3%, mostly for land transport
- Gas: 20.9%, for thermoelectric plants
- Renewables: 13.1%
- Nuclear: 6.5%

The 13.1% represented by renewable energies comes from:

-   Biomass: 10.4%
-   Hydroelectric: 2.2%
-   Other energies: 0.5%

The 0.5% of world consumption from "other" energy sources consists of

-   Geothermal: 0.41%
-   Wind: 0.064%
-   Solar: 0.039%

World production of electrical energy alone (as opposed to the total consumption of all types of energy) is as follows:

-   Coal: 40%
-   Oil: 7%
-   Gas: 20%
-   Hydroelectric: 16%
-   Nuclear: 16%
-   Renewables: 1%

By 2040, experts predict the following consumption:

-   Coal: 22%
-   Oil: 27%
-   Gas: 25%
-   Nuclear: 6%
-   Renewables: 20%

Of the world's 20% renewables, China has 11% and the US 9%.

These numbers clearly show the excessive dependence on coal and natural gas for electricity production in 2019 and, therefore, the rise in greenhouse gas emissions along with population growth. In 2019, renewable energies, mainly wind and solar, still represented only a tiny percentage of the planet's electricity generation, but these two technologies receive generous incentives and subsidies from many governments and investors around the world and thus show good growth. As can be seen above, renewables are expected to see growth of 20 times their current usage by 2040. However, considering that 67% (see above) of the world's electricity generation still depends on fossil resources, it is obvious that all wind and solar installations will never be able to replace fossil-based energy sources. Meanwhile, the growing demand for electrical energy must be satisfied with nuclear energy, instead of installing new thermoelectric plants. This technology can be considered renewable, because uranium and other fissile radionuclides exist in great abundance. Although nuclear power plants do not emit greenhouse gases, nuclear **fission** generates very dangerous radioactive waste and thus it is not a clean process. The world eagerly awaits successful results from the many developments in nuclear **fusion** around the world, which should produce completely clean electricity in abundance.

III

GLOBAL WARMING ---> CLIMATE CHANGE

This series of consecutive chapters will explain Global Warming. Its main causes are electricity generation and transportation, which are still > 70% dependent on the burning of fossil fuels, with their increasing greenhouse gas emissions. All living things on Earth are already increasingly suffering consequences that, without smart measures, could make our Earth uninhabitable within ~ 50 years. The various remedies currently under development will be insufficient until the energy we use is available in abundance without the emission of greenhouse gases. Now is the crucial time to organize a worldwide united effort to finally realize the eternal dream of physicists to generate electrical power through nuclear fusion. This process duplicates the process the Sun has used to generate its enormous energy for millions of years, and which it will continue to produce for many more without emissions, radioactive waste or danger of explosion.

1) WHAT IS GLOBAL WARMING?

Global warming is caused by the accumulation of greenhouse gases, mainly carbon dioxide (CO_2), in the Earth's atmosphere. Visible light (short wave) from the Sun easily passes through the atmosphere and heats the Earth's surface, but the increase in CO_2 does not let enough heat (infrared light) escape into space. CO_2 reflects heat back to the Earth's surface, causing the planet to heat up. This is the so-called greenhouse effect that is causing global warming.

Methane gas (CH_4) is over 20 times more harmful than CO_2 in this regard. This gas is released into the environment due to negligence and accidents in oil and natural gas production. Agriculture and livestock also emit more and more of this gas into the atmosphere, and the warming of the soil and the oceans, where CH_4 exists in the depths as a hydrate, also releases increasing volumes of CH_4. Methane contributes an estimated 15% to global warming.

As more of these gases accumulate in the atmosphere, more heat is redirected back to the Earth's surface. Other known greenhouse gases are ozone (O_3), which protects the Earth from the sun's dangerous ultraviolet radiation, and nitrous gases (NO_x), emitted, above all, by cars, trucks and airplanes with internal combustion engines and responsible for smog and high temperatures in urban areas.

The troposphere is the very thin layer of air measuring ~ 11km (kilometers), 0.008% of the diameter of the Earth (12,756 km). Airplanes fly in the troposphere, and it affects the climate. The correct CO_2 concentration of <300ppm (parts per million) ensures that the proper temperature, an average ~ 14°C (centigrade), is maintained on Earth; without CO_2 the temperature would be -27°C. But if human activity increases this level well above 400ppm, the warming will become irreversible and the survival of humanity will be in question.

Before the industrial revolution, humanity depended solely on the Sun for its energy. In very simple terms: Plants use the Sun's energy and "breathe" by absorbing CO_2 from the air and converting carbon into biomass with the chlorophyll catalyst. They then "breathe out" oxygen: that's photosynthesis (see the last chapters of this book). Animals and humans use this biomass (sugars, starch and protein) as food and, breathing oxygen, "burn" the nutrients with the help of the hemoglobin catalyst and convert them into energy, exhaling CO_2. Humans applied their own and their animals' energy for transportation, used cellulose from biomass for clothing, and wood to build and heat their shelters. All these activities, plus natural effects, such as volcanism, in no way affected their environment and the CO_2 in the air remained at 285ppm.

With the onset of the industrial age, humans began burning fossil resources to generate energy, releasing roughly twice the amount of CO_2 into the atmosphere. For the first hundred years, until about 1900, the increase in CO_2 to 300ppm was hardly noticed, but since 2000, the concentration of 400ppm is manifesting itself with more strength every year in dangerous climatic effects, because nature, trees and seas can no longer absorb the increasing amounts of CO_2.

With the Industrial Revolution (~ 1850), the Earth's population grew substantially. After the Second World War (1945), the planet still had less than two billion people. Today (2021), the world population is about 7.7 billion and it is expected to grow to 9.7 billion within the next 30 years. Likewise, human activity, such as the combustion of fossil materials, has increased exponentially, so that, proportionally, CO_2 levels have risen to 400 ppm, an increase of 40% over the preindustrial level of 285 ppm. Consequently, since 1990, the temperature of the Earth's surface has risen 1C°. Until the last 30 years, the Earth has been able to maintain a fragile climate balance with CO_2 levels of 300ppm, by absorbing much of this gas in large forests and oceans; however, systematic deforestation is worsening the situation.

The warming trend is not uniform across the planet, as the effect is more drastic in the arctic regions. Glaciers and white snow act as a mirror and reflect more than 90% of sunlight back into space without heating these surfaces, which have a very high albedo (reflection index). This is not the case for open dark blue waters, which absorb more than 80% of incoming light due to their low albedo. An albedo of 1 means complete reflection of light, e.g., a mirror; an albedo of 0 indicates full absorption, as by a completely black surface. Thus, polar ocean temperatures have risen by an average of > 2°C in the last ten years and sea levels worldwide are rising, because the melting of the ice has caused the thickness of the glaciers in Greenland and in Antarctica to decrease by 40%. Therefore, global warming is amplified in the polar regions, which are the primary systems for controlling the planet's climate.

These drastic changes in temperatures in the polar regions extend over our entire planet. For example, the great ocean currents, such as the Humboldt and the Gulf of Mexico current, are already decreasing by 15%. As a result, Northern Europe receives less heat, leading to colder winters.

Fourteen of the 2019 natural disasters, which were undeniably caused by climate change, cost in excess of a trillion dollars each and countless lives lost, not counting the recent catastrophic fires in Australia. These, plus the large fires in California and in most of the West of the United States, and the systematic burning of the Amazon, put more than a trillion tons of CO_2 in the atmosphere in 2019, over 20% more than in previous years. In 2003, heat waves, which periodically return, were the cause of over 25,000 deaths in Europe.

The global measurement of CO_2 concentration is very accurate; the most reliable numbers are from NASA, which indicates that today the level is at 400ppm and rising. Before the Industrial Revolution (~ 1800), the global level of CO_2 was ~ 280ppm, which was determined by examining air bubbles trapped in the ice extracted from deep drilling of the great ice caps in Antarctica. It is interesting to note that these air bubbles were found to contain ash from the eruption of Mount Vesuvius in AD 79.

For a better perspective on this, let's consider that CO_2 emissions from human activity in 2019 were 3.1% higher than in 2018, compared to an increase of 1.4% per year in previous years, and the population is obviously growing faster than energy sources. Global energy consumption is anticipated to grow 44% in the next two decades, primarily in India and China. Thus, by 2040, the increase in emissions will more than double. Despite commitments made in recent international conferences (Paris agreements), CO_2 emissions, on a global basis, show a continuous increase, because ~ 70% of electrical energy still depends on fossil fuels, such as coal, oil and natural gas.

The drastic annual increase in CO_2 is clearly caused by human activity. Before the Industrial Revolution, the level of CO_2 due to fires, volcanic eruptions, etc., was kept well below 300ppm by various natural processes, such as photosynthesis in the great forests of the world and absorption in the sea, where mollusks use CO_2 to form their calcareous shells and large coral reefs. Clearly, human activity is destroying this delicate balance.

Our two neighboring planets are frightening examples of what could happen on Earth. Venus, the second planet in the solar system, is twice as far from the Sun as Mercury, and still is the hottest planet, with an average temperature of 461^0C. Venus is very similar to our Earth, 95% of its size and 85% of its mass. It has a rocky surface, a nickel/iron core, and had active volcanoes. Several billion years ago, Venus was colder than it is now, but the accumulation of CO_2 produced its very dense atmosphere of 96% of this greenhouse gas, which does not let the heat that sunlight generates on the planet's surface escape due to its very low albedo.

On our other side, Mars is almost twice the distance from the Sun than Earth. It is half our size and, because it has no metallic core, it has only 10% of Earth's mass and 38% less gravity. Therefore, most of its water and air have escaped into space. With its very thin CO_2 atmosphere, Mars reflects most of the energy that comes from the Sun back into space, and thus it has a very low temperature of between -27⁰ to -46⁰ C. Mars is the opposite of Venus, whereas Earth is in an ideal intermediate position.

2) HOW THE CLIMATE DEVELOPED

Paleoclimatology explains how, since its formation some 4.5 billion years ago, the Earth has passed through many irregular alternations between hot and cold geological eras. Climate is a consequence of atmospheric development. The first 500 million years, when the Earth was still an incandescent core, were marked by incessant volcanism and meteorite bombardment. Over time, the earth's surface solidified and gases and vapors formed a primitive atmosphere. This primordial environment consisted of gases from volcanic emanations, predominantly nitrogen and CO_2, but no oxygen. Gradual cooling allowed the formation of steam clouds, giving rise to abundant precipitation that formed the first oceans. It is likely that these large amounts of water came from outside of our solar system, in the form of comets. These humid conditions allowed the first life forms to emerge, some 3.6 billion years ago.

In a significant jump, there appeared cyanobacteria, capable of photosynthesis, which absorbed CO_2 and released oxygen (around 20%); the greenhouse effect of CO_2 was reduced and the Earth gradually cooled down. Next, the planet went through its first glacial phase, caused by the impact of a large meteorite. The resulting dense layer of dust did not allow sunlight to pass through, causing global cooling. It is remarkable to see how the Earth's climate depends so much on maintaining the favorable average temperature of 14⁰C: A fall of only 6⁰C results in a "White Earth," completely covered in ice, while a 5⁰C rise has already resulted in massive extinctions several times. The division of the supercontinent Gondwana and tectonic movements caused enormous episodes of volcanism, leading to alternating periods of heat and cold through the geological ages, from the Precambrian to the Carboniferous epoch, some 300 million years ago. During these very long intervals, the planet went through multiple extinctions, but life always recovered from small living things in the depths of tropical seas.

The Carboniferous epoch was accompanied, once again, by large volcanic eruptions and the resulting high levels of CO_2 and heat, which produced lush vegetation that allowed all varieties of life to develop in gigantic forms. In the next extinction, all this organic material was covered with the erosion of the newly raised high mountains, turning everything into the fossil deposits we are burning today in excessive quantities. At the end of the Cretaceous, 66 million years ago, the last mass extinction occurred due to the impact of a meteorite in Chicxulub, Yucatán. This global catastrophe not only wiped out the dinosaurs, but was lethal for 75% of the plant and animal species that populated the Earth.

With the end of the fragmentation of Pangea into the various continents, the surface and deep sea currents were formed, becoming the most important modulators of Earth's climate. Benign environmental conditions favored the first human settlements and Mesopotamian civilizations between 10,000 and 5,000 years ago. Volcanic explosions, like the one on Santorini, also occurred during the biblical era. These catastrophes continued during modern times with Tambora, Krakatoa, Pinatubo and other volcanoes. All these episodes affected the global climate. In many cases, the drop in temperature resulted in severe famines in different regions of the globe.

During the 20th century, the climate has remained warm. Since 1950, a continuous increase in temperature has been noted, with increasing levels of CO_2 in the atmosphere. For the first time in history, the cause of such rapid change is the consequence of the actions of a species, humans, and the uncontrolled burning of fossil fuels. Thus, geologists began to call our era the Anthropocene to refer to the time when human beings have begun to affect the climate, unfortunately to our disadvantage.

3) HOW IT AFFECTS HUMANS

It is worth taking a moment to consider the significant role the weather plays in our lives, affecting both our physical and mental health. The cold months bring diseases such as flu, influenza, etc., and the dark winters in the Nordic countries are causing an increase of suicides. At the same time, extreme heat (> 40ºC) kills many people every year in European countries that have historically enjoyed a temperate climate. Environmental temperatures define how we dress and how and where we protect ourselves from extreme weather, i.e. rain or snow, tornadoes, etc. The climate also regulates our nutrition: We depend on what agriculture can produce during different times of the year, if the weather gave good or bad harvests and, consequently, how much we pay for our food. Our finances are equally affected by how much we pay for the energy that we use to enjoy, or not, a comfortable environment in our homes. Changing rainfall distribution around the world is forcing many people to flee the progressive desertification of locations in Africa and Central America. In coastal areas, hurricanes, more frequent and more devastating than ever before, cause people to abandon beach regions. In addition, rising ocean levels will motivate climate refugees to move further inland.

Earth's fauna and flora are even more affected than humans by climate change, which brings new diseases and has even brought several species to the brink of extinction. It is a fact that humanity cannot control the climate, but we have a great responsibility to correct the damage we have already caused.

4) MAIN CAUSES OF GLOBAL WARMING

To better understand increasing CO_2 emissions, we must distinguish between the two main sources of the problem: a) electricity generation in thermoelectric plants. b) the use of petroleum derivatives for transportation, be it by land, sea or air. c) Natural processes.

The production of raw materials for power plants, such as coal, oil, gas, and fuels for different types of transport, also requires energy that is generated mainly through processes that produce greenhouse gases. In the following chapters, we will examine how much energy has to be spent on the production, refining, distribution, etc. of oil, and the environmental effects and corresponding CO emissions. The definition of energy efficiency involves determining which form of energy used in the different power plants or means of transportation produces the best performance and, consequently, the least environmental pollution.

– Coal mining

Underground mining has always been a dangerous job and has had a high and deplorable toll on human lives through floods, landslides and explosions. For this reason, in the United States, coal mining is done primarily in open pit mines, where entire mountain tops are destroyed to gain access to the coal strata. This practice has caused irreparable environmental damage.

– Environmental effect of coal

Coal, oil, and gas are the products of the decomposition of ancient plant and animal life, formed in geological processes over 60 million years after the Carboniferous period. These materials became hydrocarbons, chemical compounds that contain only hydrogen and carbon atoms. Energy from the Sun produced these fossil fuels, which are still the largest source of energy today. Unlike gas, which is mainly pure methane, coal and oil, due to their natural origin, contain derivatives of sulfur, nitrogen, phosphorus, etc., even traces of toxic heavy metals such as mercury and lead. The typical medium-sized 500MW coal-fired thermoelectric plant emits annually, in addition to CO_2, 10,200 tons (t) of nitrous gases (NO_x), 10,000t of sulfur dioxide (SO_2), 720t of carbon monoxide, 220t of hydrocarbons, 110kg of arsenic, 77kg of mercury, 59kg of lead, plus 125,000t of ash, all equally toxic. Thus, this fuel is not only responsible for the greenhouse effect of CO_2, but also causes severe air pollution and diseases, especially in large urban centers.

a) THERMOELECTRIC PLANTS

Today, large thermoelectric plants satisfy the growing demand for electrical energy. Globally, coal still produces more than 45% of electricity demand but, little by little, natural gas is replacing coal, reducing CO_2 production by 40%. Just to put this into perspective, burning one ton of coal produces ~ 2.5 tons of CO_2 and ~ 7,600kWh (kilowatt-hours), which can supply electricity for a house for a month. A typical 500MW thermoelectric plant, which can generate electricity for 8 small towns, burns 1.4 million tons of coal per year and emits 3.3 million tons of CO_2, producing 3.5 billion kWh. In 2020, several countries are still building thermoelectric plants for burning coal. Coal consumption for this year is forecast at 5.636 billion tons (~ 14.0 billion tons CO_2), with China being responsible for 45%. The effect of the generation of clean and renewable electricity is still hardly noticeable (1%).

Most of the thermoelectric plants that burn mineral coal have a capacity of 500MW. This type of plant is the fastest and cheapest to build. Its raw material, coal, is even the most economical fuel. The useful life of a thermoelectric plant is ~ 20 years. These plants are located in regions with urban centers to produce electricity where it is needed and thus avoid costs and losses from transmitting and distributing electricity over long distances. With increasing attention to global CO_2 damage and local effects (air pollution), more modern plants use various modifications to reduce the emission of toxic smoke. But there is no inexpensive technology for reducing CO_2. The large amounts of toxic ash also await an ecological solution. Ten plants in the world have 5,500MW capacity. A typical plant transforms only 30% of the heat produced by burning coal into electrical energy, since much of the heat is released into the atmosphere in the combustion gases and another part of the residual heat from the steam from the electric turbines is absorbed in the cooling water used at such plants, warming rivers. To reduce this wasted energy, cogeneration plants use this heat in other industrial and chemical processes. Energy efficiency can thus be increased up to 70%.

Recently, China has dedicated large financial and scientific resources to reduce the damage caused by its large thermoelectric plants, which still depend on coal, and thus meet its goals of reducing its CO_2 emissions. China's newest plant has very complex and expensive equipment that captures 90% of the CO_2 in its combustion gases. The gas passes through multiple stages of purification and chemical processes, and the pure CO_2 is compressed and injected as a liquid into rock at great depth. The government justified the high costs of this facility as a prototype for cheaper future facilities. This plant avoids the emission of 150,000t of CO_2 into the atmosphere per year.

In 2020, billionaire Elon Musk, CEO of Tesla, established a prize of $100 million for an economic process for the capture of CO_2.

— Natural gas (methane, CH_4)

Methane is the cleanest fuel. More and more new thermoelectric facilities use natural gas and thus cause less pollution than centers that still burn coal or fuel oil (an estimated 60%). Natural gas does not produce nitrogen oxides or sulfur dioxide and generates ~ 30% less CO_2 than oil and ~ 40% less than coal. But in boilers or gas turbines used to generate electricity, its thermal efficiency maxes out at 60%. These plants depend on pipelines for gas supply, because liquid gas requires transport at very low temperatures. The utility of these facilities is very high. They need little maintenance and in event of a failure of the gas supply, they can easily be adapted to the use of diesel or LP gas (liquid hydrocarbons of 3-5 carbons).

In some regions of the world, due to the urgency of getting quick access to electricity or where there are no high voltage transmission lines, such as in Baja California Sur and in the Yucatán, thermoelectric plants are still being built. Under such circumstances, the least bad option for the environment are plants that use natural gas, where this fuel is widely available through pipelines. Affordable transportation of natural gas depends on these pipelines, because to convert methane into a liquid at atmospheric pressure, its temperature has to be lowered to -162°C, which reduces its volume up to 600 times. Liquid natural gas (LNG), produced in remote isolated areas, is transported in large gas tankers, called LNG carriers, and then regasified for use. Gas-fired thermoelectric plants are still the most economical, because the costly storage and handling of coal or fuel oil is eliminated. The energy efficiency of thermoelectric plants is 30% with coal and up to 60% with gas.

Natural gas is widely available in the Americas and in other regions of the world. Fracking is the hydraulic fracturing and extraction of gas and light oil from shale formations from one thousand to five thousand meters below ground. This technique made the United States self-sufficient in fossil fuels and turned the country into a net oil exporter. This process requires strict environmental protections and, for this reason, it is prohibited in Mexico and several other countries around the world.

b) TRAFFIC AND TRANSPORTATION

– Oil

Today, in 2021, more than 90% of traffic and transport still depends on hydrocarbons, whose biogenetic origin (created biologically) has been proven, because crude oil has an abundant amount of porphyrins in its composition, a product of the degradation of blood and of chlorophyll. In the depths of the Earth and below the bottom of the oceans, oil and gas migrated to reservoirs between impermeable rock forms, from which they can be recovered through the drilling of wells.

In most of the world's oil wells, oil no longer reaches the surface by its own internal pressure underground and so must be pumped. This is a very inefficient process, because only a third of a reservoir is recovered. Tertiary production, such as the injection of nitrogen, CO_2 or steam, or chemical recovery, using solvents, among other methods, can bring up to three-quarters of the total reserve to the surface. These techniques are much more complicated and require greater energy expenditure. Oil production has, on average, the following energy efficiency: the energy of one barrel of oil is needed to recover up to 30 barrels from conventional wells, while with tertiary recovery, five times more energy is required.

The international market defines the price of oil in US dollars (US $). Due to serious problems from the recent pandemic, this price has been very volatile, varying between US $30 to US $70 per barrel. This unit is used as a measure in the oil industry and, depending on the density, it contains between 119 - 151 kg of crude oil or 160 liters. The cost of production depends on the technique used and to a great extent on the country. Arab countries produce a barrel of crude at a cost of US $ 3 to US $ 6, because their wells are mainly onshore. Offshore recovery costs US $ 20-30 and oil from ultra-deep offshore wells (> 5000m) costs US $ 48-52. Fracking, which mainly generates natural gas, produces very light oil at a cost of ~ US $ 30 per barrel. In Mexico, Pemex produces a barrel of crude at US$ 14.2. The quality of crude oil also determines its price. The refining of a light oil, such as Brent from the North Sea, results in better gasoline and diesel production and thus, normally, it obtains a price of up to US$ 10 higher than heavy oil, such as crude from Venezuela and Mexico.

Crude oil is refined and subjected to various thermal and catalytic processes in order to produce, above all, gasoline and diesel for land transport, jet fuel for aviation, a 16-25 carbon heating liquid, and the heaviest fuel oil used for maritime transport. Asphalt is the residue from distillation. The catalytic refining of heavy oil cuts large hydrocarbon molecules into shorter units, producing gasoline with 5-12 and diesel with 12-20 carbon atoms. This process consumes a lot of energy and thus generates high CO_2 emissions. Energy costs represent 40% of the total cost of the end product.

A standard 160-liter barrel of oil typically produces 76 liters of gasoline and 26 liters of diesel, a yield of 63.75%. Compared to any other fuel, gasoline and diesel represent the most compact and portable source of energy, and they also have the highest energy content (except for uranium). This facilitates their transportation, an essential aspect for aviation. Oil and gas are most practical for space heating in homes and businesses and, in the United States, still account for 30% of total hydrocarbon use. All these applications generate triple the amount of CO_2 (1kg diesel -> 3.16kg CO_2) and thus constitute major contributors to global warming.

– Transportation using gasoline and diesel

Land, sea and aviation traffic that burns petroleum derivatives, such as gasoline, diesel, etc., produces 60% more CO_2 than the production of electricity in thermoelectric plants. Measures to reduce air pollution have changed the global automotive landscape. The different regulatory bodies increasingly squeezed manufacturers, forcing them to produce more efficient cars to minimize harmful emissions: mainly CO_2, CO and nitrous gases (NO_x), which are responsible for smog and unhealthy air in urban centers.

Normally, the efficiency of a vehicle is defined by its fuel consumption, but many other factors, such as aerodynamics, weight and power determine this consumption. In cars with combustion engines, the energy losses are high; the average energy efficiency of cars using gasoline is only 20-25%, and trucks using diesel reach 30-35%. In hybrids, this efficiency is almost double (45%). The internal friction of combustion and traction engines (transmission, wheels, tires), the essential cooling (water or air) of the engine, and the heat of the exhaust make this waste of energy inevitable. Thus, transportation emits more CO_2 than any other human activity, more than the generation of electrical energy. A liter of gasoline produces 2.32kg of CO_2 and a liter of diesel a little more, 2.65kg of CO_2.

The growing number of electric cars for personal use do not produce emissions. Their energy efficiency reaches up to 75%. Thanks to their fewer moving parts, they present less energy loss due to friction. But the capacity of solar panels and wind turbines is still insufficient for meeting demand from the growing fleet of electric vehicles. Recharging these cars with energy from the best thermoelectric plants (using gas with a 60% efficiency) results in a total energy efficiency of only 42% (70x60). In conclusion: Electric cars present a benefit for the environment only when the electrical energy they consume comes from renewable resources.

In aviation, large intercontinental aircraft with jet engines have a maximum energy efficiency of ~ 20%, but are the most economical with respect to CO_2 produced per passenger transported. They fly at an altitude of ~ 10,000 meters and thus deposit the CO_2 from their exhaust directly into the troposphere (10km), where the climate develops. Although ultralight aircraft have been tested, some with solar panels and others with electric propulsion, it is difficult to think that this technology could be used in commercial aviation.

– Transportation using natural gas

The most suitable gases for transport, such as methane (CH_4) and hydrogen (H_2), only exist as a liquid at extremely low temperatures: CH_4 at -161^0C, H_2 at -252^0C; the volume of the gas is 600 times greater than the volume of the liquid and compression alone does not convert these gases into liquids. For this reason, its use in transport requires tanks resistant to high pressure, large enough to offer an acceptable driving range. In Argentina, a limited number of personal cars use natural gas and provide a range of more or less 100km. Buses with natural gas are also used in some urban centers, but very few stations supply this gas at sufficient pressure. The emissions from these vehicles are very clean, but the amount of CO_2 per km is very similar to the use of gasoline.

– Transport using hydrogen

Clean hydrogen, so far, is only produced via water electrolysis (HO_2), with an energy efficiency of 65%. The process requires a lot of electrical energy because the bond between hydrogen and oxygen atoms (HOH) is extremely strong. But it yields a lot of energy when hydrogen is burned with air and produces only water vapor. At first glance, hydrogen seems to be the ideal fuel, because it contains three times more energy per unit mass than gasoline. The problem with hydrogen, however, is its volume, since it cannot be used in liquid form (only at a very low temperature of -252.87⁰C). A tank in a normal car can hold 5-6kg of hydrogen under high pressure, with a range of ~ 60 miles per kilo. The use of hydrogen for transport requires special precautions, because the gas accelerates the cracking of steel (hydrogen fracture). For this reason, its transport requires a special coating for high-pressure cylinders. Hydrogen leaks are especially dangerous, since it is the only known gas that heats up and can reach its point of ignition when depressurizing. The few hydrogen cars that exist today use fuel cells (to be discussed in a later chapter), allowing their total energy efficiency to reach 40%.

The industrial process to produce hydrogen is the reaction of methane with water at temperatures > 1000⁰C, ($CH_4 + H_2O$ à $CO + 3H_2$, followed by $CO + H_2O \rightarrow CO_2 + H_2$); the process requires high pressures and temperatures and thus produces a lot of CO_2. Today, 70 million tons of hydrogen are still produced by this reaction, mainly to produce ammonia (NH_3) for fertilizers. This process does not generate gas for vehicle use, for both economic and environmental considerations.

The world is waiting for a nuclear process, fission or fusion, that allows the production of hydrogen in an environmentally acceptable and economical manner.

In conclusion, when sufficient clean electrical energy is available from renewable resources that do not produce greenhouse gases, vehicle traffic should use only electric propulsion. However, land transportation contributes only 16% to climate change. This makes it urgent to develop other sources of clean electrical energy for all other uses.

– Biofuels

The 1973 oil crisis, caused by political pressure from OPEC (Organization of Petroleum Exporting Countries) and the resulting extremely high prices (US $ 140/barrel), motivated many countries to urgently seek alternative sources of energy. In addition, the gradual decline in global oil production showed that world reserves are finite. Fracking eased these concerns for at least another 50 years.

Through photosynthesis, agriculture produces plants with high energy content, such as carbohydrates, sugars and their polymerization products, such as starch and cellulose. Parts of these carbohydrates are converted into oils, stored in seeds, such as corn, soybeans, and palm nuts, to produce energy necessary for germination. Chemical processes convert carbohydrates into ethanol and oils into diesel. Today, a relatively peaceful political environment, low oil costs (US $ 30-40), and excessive production create an economic disadvantage for biofuels, except for bio-alcohol. International criticism has also grown over the production of biofuels from food crops due to the environmental impact of deforestation to gain more land for cultivation, and because of global food prices. Depending on their energy content, biofuels generate CO_2 in amounts comparable to fossil fuels, but they are generally cleaner and thus cause less smog.

The only biofuel produced today in large quantities is ethanol from sugar cane that grows in tropical and subtropical regions, where it produces three harvests every two years. One sowing lasts up to 20 years. Processing is a simple fermentation of the extracted syrup and its distillation. Taking into consideration all the steps from planting to distribution, sugarcane ethanol has an energy balance of > 3, which means the energy produced is more than three times greater than the energy expended. In addition, the solids from the used cane, called bagasse, can be burned to produce a large part of the energy required in production.

Biofuels from foods such as corn, soybeans, etc., are rarely produced anymore because they require more expensive chemical processes and their energy balance barely reaches between 1.3 to 1.5, which is economically unviable.

The ethanol molecule (C_2H_6O) contains one oxygen atom, which means that it has only about 70% the energy content of gasoline. The combustion of 1kg of ethanol produces 1.91kg of CO_2. In Brazil, the leading ethanol producer, most personal cars have already used ethanol for many years. These vehicles use 30% more ethanol than gasoline for the same mileage. The resulting CO_2 emission is the same as for gasoline, but the combustion in the engines occurs at lower temperatures. Thus, the nitrogen in the air is not involved in combustion and the exhaust is free of nitrous gases. This is the main reason why the atmosphere in the large urban centers of Brazil is cleaner than that of other countries that use gasoline in their cars.

The advantage of ethanol is that it has a higher octane value of 113 versus 87 for gasoline, making it possible to increase the engine's compression ratio for greater thermal efficiency. For this reason, many states in the US sell blends of 15% ethanol in gasoline to reduce air pollution.

c) NATURAL PROCESSES

Several natural processes produce greenhouse gases, mainly methane: the decomposition of organic compounds from garbage dumps and drainage channels, and waste from the production of paper and sugar. This biogas is formed by the anaerobic digestion (without oxygen) of organic material. This has no great effect on climate change, despite the fact that methane has an effect 24 times stronger than CO_2. But large herds of cows and other herbivorous animals, which emit methane as part of their digestive process, contribute significantly to climate change, because a cow generates an average of 100kg of methane per year. Thus, the growing production of meat and dairy products is a threat that worsens global warming, considering that Brazil alone has 200 million cows.

d) CONCLUSION

All leading climatologists agree that if we do not manage to stop the increase in CO_2 emissions, the global temperature will increase more than 1.5°C within 10-20 years, with catastrophic consequences: irreversible warming, total drought in the equatorial regions, a sea level rise of 1.5-2 meters, and millions of climate refugees from coastal areas. A rise of 1.5° C is the maximum that the biosphere can still tolerate; 3° C more, as projected for the end of the century, will make life on Earth impossible.

Large insurance companies no longer issue insurance policies for private properties or hotels in coastal areas closer than one kilometer from the sea. Fortunately, large companies do not doubt climate change. However, many of the world's politicians prefer to ignore the danger of global warming because measures in favor of decarbonization can affect their country's economy and put their own position at risk. Without drastic measures, which are not yet well defined, the temperature will rise 3-5° C by the end of the century and the Earth may become uninhabitable.

The United States Department of Energy recently concluded that in 2050, 40% of the world's energy will still be generated in thermoelectric plants burning coal or natural gas because, despite all the efforts and subsidies, clean renewable energy from solar panels and wind turbines cannot grow as quickly as the world population. Thus, the future of our planet depends solely on the success of nuclear processes that do not produce CO_2.

5) CONSEQUENCES OF GLOBAL WARMING

The effects of global warming include environmental, social, economic and health changes; some are localized, others global, a few are reversible, some are not, and most have adverse effects. Only a few have temporary positive consequences, such as agriculture in regions further north. For example, large areas in northern Canada now produce corn, where it was previously impossible, because corn requires six months to grow for a good harvest.

Today, climate change is evident in many parts of the world. NASA has shown, with very clear maps, the drastic reduction of ice in the Arctic and Antarctic regions. The huge Greenland ice sheet receded so much that the Danes brought their cows to graze in the green littoral areas. All the glaciers in the world are in retreat and some have already disappeared completely: There are no more glaciers in "Glacier National Park" in the state of Montana in the United States. Even in the Himalayas, the thaw is evident. The effect of warming is more drastic in arctic areas. Although more than 90% of the sunlight reaching the reflective white surfaces of glaciers returns to space, the dark waters of the open sea absorb more than 80% of the light, resulting in higher water temperatures, which accelerate the melting of the glaciers.

Many islands, especially in the Oceania region, are disappearing under the rising sea waters. A total of 136 coastal cities, with 400 million people, such as New York, Miami, Tokyo, Bangkok, Venice, etc., already have problems during high tides. At the turn of the century, when the sea rises 1.5 meters further, it will force millions of climate refugees to leave their cities. At the other extreme, equatorial regions, such as Central Africa, are suffering desertification with prolonged droughts. From Central America, large migrant caravans are moving north through Mexico because of increasingly severe water shortages and related political problems. A NASA map shows that 35% of the surface of the continents can be considered desert areas. Global warming is causing extreme weather events, such as more intense and more frequent wildfires, tornadoes and hurricanes.

One aggravating factor is that, in Nordic regions, the "permafrost" (permanently frozen subsoil) is melting and the methane (> 20 times more harmful than CO_2) trapped below the surface for millennia is escaping into the atmosphere. The oceans are still the most important sink for CO_2, but its increase causes acidification by the formation of more carbonic acid. In this overly acidic environment, mollusks cannot form their solid calcareous (calcium carbonate) shells; so the great coral reefs are dying. The melting of polar ice causes an increasing flow of very cold fresh water from the polar regions, producing changes in ocean currents and the stability of the Gulf Stream. Large ocean currents have the greatest influence over Earth's climate.

Climate change is gradually pushing agriculture further north, while southern regions are suffering more droughts. These are just a few very specific examples of global warming. Today, the effects are much more disastrous. The consequences for humans include displacement of populations and pests that will move north and cause diseases, such as malaria and other tropical ailments, affecting people in regions that are still free of them. Flora and fauna are threatened by a general loss of biodiversity. Changes in agricultural productivity increase food prices and constitute a risk to food security in vulnerable populations. Agriculture is limited by water shortages and an increase in new pests. Like humans, livestock are also suffering from new and worse pests, and fisheries are declining due to the loss of reefs, which are the cradles of the sea, providing food for many maritime species.

The main political measures for facing climate change are the reduction of greenhouse gas emissions and adaptation to their consequences. The big problem is that even if we achieve a reduction in CO_2, methane and nitrous gases, it will take decades or centuries until a beneficial effect is shown. The cost of climate change, on a global basis, is a reduction in world GDP (gross domestic product) of up to 20%. It is not just the pessimists who are forecasting climate wars due to lack of water and limited soil viability.

IV

RENEWABLE ENERGIES

The following is a summary of temporary solutions to global warming, with details of the solutions available today.

As of today, no adequate and affordable remedy exists to completely eliminate the CO_2 emissions produced by electrical power generation, and global warming cannot be controlled until there is sufficient and clean electrical energy available. Even after that, it could take several decades until a suitable climatic balance is recovered.

The world's great hopes are solar panels and wind turbines to eliminate the use of fossil fuels that produce CO_2. However, solar panels depend on sunlight, available an average of 8 hours a day in the most suitable places. And eolic generators rely on wind, which normally dies down at night. In addition, there is a shortage of long-life, high-charge batteries to store excess energy generated during the day. Solar panels can be installed in desert locations, but modern windmills need large areas where other activities are limited, and their installation at sea costs more than twice as much. An important factor in the installation of equipment to generate renewable energy is the energy density on the ground. One solar panel requires 1^2m and produces 20W, and a wind turbine produces only between 3.5 to 5MW; thus, much more land is needed to replace a normal 500MW thermoelectric plant. The intermittent availability of clean and renewable energies requires thermoelectric plants to be available to compensate for the deficit and ensure a continuous supply of electricity. Gas plants are the most suitable, because they can be started quickly and adjust to demand.

The energy from the latest solar panels can already compete with the cost of electrical energy from thermoelectric plants, but wind energy is still more expensive. If a normal wind turbine produces only a maximum of 5MW and a normal thermoelectric plant produces 500MW, then more than100 wind turbines are needed to produce the same amount of electrical energy. Another important consideration is that the tall towers and foundations required for wind turbines require hundreds of tons of steel and cement and the manufacture of these materials and their transport produces many tons of CO_2, as explained in the following chapters.

Even worse, environmental experts have concluded that even with all the possible solar panels and wind turbines, renewable energy can only reach a maximum of 30% of the global demand for electricity. Today, this deficit can only be met with fossil fuels or nuclear power plants. For this reason, CO_2 emissions continue to increase along with population growth, estimated at 44% over the coming decades. It is obvious that the population is growing much faster than renewable energy sources.

Other sources of non-polluting and renewable energy are hydroelectric and geothermal energy. Most of the great rivers of the world already have modern dams, but in many, climate change is already manifesting in a lack of water. Despite the fact that geothermal energy produces almost no greenhouse gases, the Geothermal Energy Association estimates that, to date, only 6.5% of its global geothermal potential is being used.

1) AVAILABLE ALTERNATIVE ENERGIES

–   Geothermal energy

This renewable energy takes advantage of the heat of the Earth: Below the Earth's thin crust, also called the lithosphere, which lies only 7 km below the oceans and 70km below the continents, is the so-called terrestrial mantle, which consists of magma, a viscous mass of molten rocks at a temperature of 3000⁰C. The thermal gradient defines the variation of temperature with a given depth. In general, the temperature rises 30⁰C per kilometer of depth. In many regions of the Earth, especially in areas where tectonic plates collide, there are areas where magma is closer to the surface. In some places, it breaks through the crust and forms volcanos.

These thermal fields are home to geysers, hot springs and volcanos. All thermoelectric plants are located in these regions, where the heat of the Earth is more easily accessible. Thermoelectric energy is considered renewable, because the heat extraction is very small compared to the internal heat of the planet.

The construction of geothermal plants is more expensive compared to thermoelectric plants because in order to reach sufficiently hot geothermal reservoirs of >150°C, at least two boreholes of up to 3000 meters deep are required: one tube to inject water and the other to release steam, which has small amounts of CO_2 and acid impurities. For this reason, metallic installations must be resistant to corrosion. Greenhouse gas emission is only 5% of thermoelectric plants. Twenty-four such plants exist In the entire world. California, in the United States, generates 7,525KW of geothermal electricity, 7.5% of the country's total demand. Mexico produces 6,040KW, which represents 1.84% of its national consumption. Only Iceland meets its entire need for electricity through geothermal and hydroelectric power.

– Hydroelectric energy

The Sun's energy produces the "water cycle," evaporating water from the seas and forming clouds, which lift humidity to the cold heights, where the water condenses and falls on the Earth as rain. Renewable hydroelectric energy, which is renewed through thawing and precipitation, takes advantage of this eternal cycle by transforming the gravitational force of water into electricity. Hydroelectric plants are built in the narrow mouths of large river canyons. Huge lakes form behind the dam, which must have a suitable height between the water level in the upper part and the turbines below the dam. This makes it possible to take advantage of the gravitational energy of the water created by this height difference. The kinetic (mechanical) force of running water rotates turbines coupled to generators to produce electricity. The fundamental source of hydroelectric energy is the Earth's gravity.

Modern hydroelectric plants are the evolution of the old hydraulic mills used to grind grains. Most modern facilities are located directly on the course of a river. In different mountainous regions of the world, smaller dams with less water flow were built at high elevations and the water flows through high-pressure tubes to turbines more than 1,000 meters below. In this case, a smaller flow, but with higher pressure, can generate energy comparable to large low-pressure plants.

Many "reversible plants" have systems that pump the water leaving the turbines back to the dam during periods of low demand on the electricity grid, further increasing the daily load factor. In places where there is sufficient height difference, this system has a great advantage because its rapid response to changes in demand can compensate for the intermittency of energy from solar panels and wind turbines.

The energy efficiency of hydroelectric plants is 90-95%. These facilities generate hydroelectric power that accounts for a fifth (~ 16%) of the world's total electricity demand and, most important, creates no greenhouse gas emissions. Hydroelectric plants require a high investment. Their construction can last several years, but they have a very long useful life and minimum operation and maintenance costs. The power of a hydroelectric plant is a function of difference that exists between the level of the reservoir and the turbines below. One great advantage is that the gate can be opened and closed to control the flow of water according to the electricity demand. Dams also serve to regulate river flow, prevent flooding, and provide irrigation. Good management of the dam level is essential to prepare for periods of high rainfall and drain excess water before precipitation exceeds the dam's safe level and causes problems for downstream populations.

Many old dams made of earth and rock are still in use, but after several tragic accidents, they are gradually being deactivated. The first hydroelectric power station was installed in 1881 in Niagara Falls between Canada and the United States. The largest centers in the world are the Great Coulee of the Columbia River, in the state of Washington, with 6,480MW; Itaipu, between Brazil and Paraguay, with 14,000MW; and, the largest, in the Three Canyons of the Yangtze River of China, with 22,500MW, which is equivalent to 45 thermoelectric plants and thus prevents 1,485 billion tons of CO_2 from being released into the atmosphere per year.

Mexico has more than 400 dams distributed throughout its territory, mainly to supply the population and for agriculture. Eighty-six of them are medium-sized hydroelectric plants with an installed capacity of 12,642MW, which generates 10% of the country's electricity demand.

The selection of hydroelectric plants requires a careful analysis of their environmental impact, as well as contemplating the loss of large areas of fertile soil, entire villages submerged under the dam lake, population displacement, and the effect on the migration of fish, such as salmon. Unfortunately, images of dams half empty or dry are common in today's world news due to changes in rainfall patterns created by climate change. Most of the great rivers of the world already have their maximum possible number of dams. Only the Congo River in Africa awaits the completion of its first dam, mainly due to political instability in the region. A new trend favors building small-scale hydroelectric plants, with capacities between 50 and 100MW, that can supply energy for a single community. Areas with abundant hydroelectric power attract important industries, such as a large aluminum plant in Iceland. The life of certain hydroelectric plants is limited, as rivers carry more and more silt that is deposited at the bottom of the dam, reducing its usefulness in the long term (e.g., the Aswan Dam in Egypt).

– Renewable energy from the ocean

A moderate current of water or a minimal ocean current (tides or waves) can be converted into considerable amounts of electrical energy. Large-scale experimental facilities are working with five very promising and proven concepts, but they still have only reached limited use:

Ocean current energy uses the kinetic energy (flow) of large ocean currents, for example the Gulf Stream; however, so far this is only an idea.

Tidal energy captures the energy of the water moved by the tides. Several submerged turbine installations already exist at the bottom of straits where large amounts of water flow every six hours, for example in the Bay of Fundy in Nova Scotia, Canada, where the difference in height between high and low tide is 18m. A facility in the Strait of Gibraltar is still awaiting funding. Other sources of ocean energy include:

Conversion of ocean thermal energy.

Wave energy that uses the energy of ocean waves.

Gradient or osmotic saline energy.

Cooling with water from deep lakes: In Chicago, several buildings control their temperature with cold water from Lake Michigan.

Ocean currents are the product of the Earth's rotation and differences in temperature. The tides, on the other hand, are caused by the gravitational forces of the Moon. Variations in the height of the tide are greater in the high latitudes of Earth's north and the south, while near the Equator these differences are minimal.

– Solar fuel

Gasoline and diesel are the two resources with the highest and most concentrated energy content. As liquids, they are easy to transport and widely available, but they are the biggest damaging factor in climate change because their use generates greenhouse gases. To prevent the advance of global warming, renewable and clean electrical energy is the most logical solution, because electricity derived from energy from the Sun, such as photovoltaic or wind energy and nuclear energy, has no negative environmental impact. A disadvantage of this electrical energy is that it cannot be stored or transported like fossil fuels, and its energy density is very low, a factor that limits its use in several essential applications of modern life, such as in aviation and areas outside the power grid, etc.

The US Department of Energy is offering to fund university and private industry projects to develop processes that convert the Sun's energy directly into easily portable and storable energy resources. This fundamental research is focused on direct processes, such as converting energy from the Sun, in a single step, into a form of energy for direct use without additional transformations. For example, the double transformation of energy from the Sun into photovoltaic electricity and subsequently, in a second step, using it in the electrolytic hydrolysis of water to produce hydrogen as fuel, has very low energy efficiency (~ 10%) and, thus, is economically unviable. Every transformation from one form of energy to another results in considerable losses.

This essential research is focusing on photochemical, electrochemical, photobiological and thermochemical processes to reduce water (H-O-H) directly into hydrogen as fuel. The goal is to weaken and break the strong bond between oxygen and hydrogen through catalysts. The thermochemical process uses high heat, concentrating the entire spectrum of sunlight through multiple concave mirrors at a single point, where temperatures > 1500°C release hydrogen and oxygen from water as gases. Photobiological reactions are also known to occur in algae and bacteria, which use sunlight to reduce water and release hydrogen.

In a solar photochemical process, a cell with a photosensitizing electrode converts light into an electric current that directly decomposes water into its O_2 and H_2 components. This dye-sensitized semiconductor solar cell uses metal oxide catalysts and directly produces H_2. Several proofs of concept have already demonstrated this reaction, but they are not yet sufficiently developed for commercial use. However, its relative simplicity gives the advantage of potential lower costs and improved energy conversion efficiency.

Hydrogen is the ideal fuel, because it is transportable as compressed gas and its burning produces high energy and only emits water vapor. This research is still in its infancy but has received a lot of interest recently, because hydrogen could become the ideal energy for electric vehicles through fuel cells.

V

STEPS TO REDUCE EMISSIONS

---

1) ELECTRIC CARS

It is possible to foresee a future in which all land transport will gradually replace the internal combustion engine (gasoline or diesel) with electric motors. For maritime and aviation transport, that process seems less feasible. In urban centers, electric cars will eliminate the toxic smog that causes serious health problems for the population in many cities around the world. However, at the moment, the renewable electricity supply is insufficient to satisfy all the needs of land transport. This deficit can only be satisfied with energy from thermoelectric plants, which relocate pollution outside of cities but do not reduce CO_2. For personal electric cars, a range of 300-400 kilometers may be adequate. For long-distance land transportation by car and, especially, by truck, better battery technology is essential.

Governments are using different incentives such as tax exemptions and various other aids to promote the use of electric cars. Regarding the environmental impact, electric cars have an efficiency of 77% if they are recharged with renewable electric energy, but they only achieve a performance of 42% if the electricity is generated in thermoelectric plants burning natural gas. It is surprising that electric vehicles are still considerably more expensive than cars with internal combustion engines, despite the fact that they use far fewer moving parts and obviously require much less maintenance.

Lithium ion batteries are the "heart" of electric cars and are also used extensively in most portable electronic equipment because of their higher charge density and because they allow hundreds of recharges over their long life. Car batteries are still very expensive; those that offer the greatest autonomy, up to 500km per charge, use materials with limited availability, such as cobalt and others. Many large firms have dedicated their best scientists and millions of dollars towards improving performance and reducing the cost of storing electrical energy. Special attention is focused on the loss of charge over time, reducing charging times, the number of recharges, safety and size, etc.

All developed countries are working to install charging stations, but there are still not enough for the general public. Normally, urban vehicles are recharged overnight at home, because a full charge takes several hours. Drivers, anxious that their battery might run out before reaching their destination, can get a quick charge in half an hour at public stations but must bear in mind that fast recharges limit battery life. For this reason, chargers are being installed in places where people leave their car for several hours, such as outside of restaurants, shopping centers, even in church parking lots. Some cars are equipped with "regenerative braking," which improves their autonomy.

2) FUEL CELL

An internal combustion engine can use hydrogen as fuel, but this gas is burned producing thermal energy that generates the mechanical energy to drive the pistons, etc. Most of the energy is lost as heat. Fuel cells transform hydrogen directly into electricity for electric car motors without generating heat. These cells are electrochemical devices in which a continuous flow of hydrogen and air produces an electric current in an external circuit. Hydrogen is supplied at the anode, which is split into protons and electrons. The protons migrate to the cell's cathode through the membrane, where they react with oxygen in the air to form water vapor.

In a hydrogen economy, where compressed gas is widely available at charging stations, a fuel cell could be used in electric vehicles with hydrogen tanks instead of heavy batteries. Charging compressed hydrogen takes a few minutes compared to several hours to recharge a battery, and the vehicle's range depends only on the size of its tank. When the energy efficiency calculations of the two processes are compared, the advantage of fuel cells are obvious.

In conclusion: Electric transport has a beneficial effect on the environment only when all the electrical energy used is produced by renewable and clean resources, without the emission of greenhouse gases.

3) IMPROVE OIL PROCESSES

The concentration of methane in the atmosphere is minimal when compared to CO_2, but it is 24 times more damaging in terms of climate change. Thus, its release in oil processes, exploration wells, production and refineries must be better controlled. Unfortunately, many maritime deposits, and several onshore deposits, still do not have the necessary infrastructure to recover natural gas, which is burned or released directly into the atmosphere. In the Mexican part of the Gulf of Mexico, for example, ~ 150,000 million cubic meters of natural gas are burned annually, releasing 400 tons of CO_2 daily into the atmosphere, not counting the methane that comes out unburned.

4) PRIVATE INITIATIVES

Some countries have committed to planting a million trees, a highly recommendable effort, but unfortunately most of these big projects forget that new plantations require good care, water, and at least ten years to mature. To achieve a notable reduction in CO_2, a trillion of large trees would be needed. Unfortunately, in the Amazon region, large areas of pristine tropical rainforest are burned to gain ground for soybean cultivation, a highly irresponsible process that produces millions of tons of CO_2 and exhausts the thin soil layer after only two harvests. This loss of the majestic trees, which absorbed so much CO_2, is irreplaceable. Climate change cannot be stopped with reforestation. Each tree planted absorbs 10-30kg CO_2 per year and more than double that figure in tropical regions; however, every small individual contribution helps.

Vehicles using petroleum derivatives and, above all, electricity generation are still the main emitters of CO_2; thus, a valuable individual contribution is minimizing the use of motorized traffic and using public transport. In our homes, many opportunities exist to reduce the consumption of electrical energy without sacrificing personal comfort by converting all lighting to LED bulbs, with a savings of > 90%, controlling indoor temperatures with timed thermostats, etc.

The recent COVID-19 pandemic has forced most business meetings to be held virtually through video-teleconferences, which proved equally efficient as personal meetings. This successful demonstration can prevent many business trips in the future and save companies money. Airlines estimate that 40% of their flights represent travel by business people or politicians. A reduction in this air traffic could prevent millions of tons of CO_2 from accumulating in the highest regions of the atmosphere.

Many private and government initiatives aim to save energy and thus reduce CO_2. Some governments have levied additional taxes on gasoline to reduce the use of personal cars. Some countries also charge a carbon tax for each ton of CO_2 emitted. A thermoelectric plant developed a process to clean, compress, separate and inject CO_2 into very deep geological formations, but the cost and risk in case of earthquakes are prohibitive. Earthquakes can release CO_2 that builds up on the surface, causing people to suffocate.

Saving water is another opportunity for conserving energy, because pumping consumes electricity and in many arid regions of the world, drinking water is produced from salty sea water through reverse osmosis. This is the most common process, but it requires extremely high pressure and thus consumes a lot of electrical energy. Several old facilities are still operating a system that boils seawater and condenses steam, a process that consumes a lot of oil. Only with clean and abundant electrical energy can we reverse the growing shortage of fresh water and realize the ancient biblical dream of making deserts green.

Consuming less red meat or dairy products can help reduce large herds of cows and thus avoid methane emissions the digestive process of these animals produces.

People have numerous individual opportunities to contribute to a sustainable economy, starting with recycling, controlling the use of plastics, etc., and using citizen power over politicians to draft better environmental laws.

VI

AREAS WHERE COAL AND OIL ARE ESSENTIAL

Several politicians, acting on misinformation, strictly recommend banning all coal and oil activities and eliminating all facilities that produce or use coal and oil products, in order to completely eliminate any possibility of emitting more CO_2 into the atmosphere and thus reduce global warming. A number of good reasons make this idea impossible, at least with the technology available today.

With the increase in the world population, and considering all the possible facilities used to generate renewable energy, the world has a global deficit of clean electrical energy that stands at 30% today and will grow to 70% by the end of the century. Only more nuclear fission reactors can satisfy this need until nuclear fusion (see the following chapters) becomes a practical reality.

1) COAL

In some metallurgical and petrochemical processes, there is no substitute for coal or other fossil fuels. Every year, the growing world population needs more construction, buildings, roads, bridges, etc., requiring enormous amounts of steel and cement. To produce 100kg of steel, you need 77kg of coal. For 100kg of cement, you need 22kg of coal. These two combustion processes inevitably emit the corresponding amount of CO_2 into the atmosphere.

Several other metallurgical processes depend on coal, the most important being the production of the element silicon by reducing its dioxide with carbon (SiO_2 + 2C àSi + 2CO). Molten quartz reacts with coal in an electric arc furnace at temperatures of 2000°C to yield metallurgical silicon of 98% purity. These extreme conditions require a lot of electrical energy and emit large amounts of CO_2 into the environment. Quartz is the most abundant material (in the form of sand) on Earth. Silicon metal is used directly for alloying with aluminum and to meet the high demand for silicones (oxygen polymerization compounds of Si) and, because of its semiconductive property, it is used in the electronics industry and in solar panels, after a long and expensive purification and crystallization process. Coal is also essential for producing mercury, tin, and lead.

It should be noted that the manufacture and installation of equipment to generate clean and renewable electrical energy, such as solar panels and, especially, wind turbines, produces large emissions of CO_2. The manufacture and installation of a wind turbine generates so much CO_2 that it would have to operate for more than thirty years for the electrical energy it produces to compensate for the environmental effect caused by its presence. A wind generator requires ~ 250t of steel, which requires 193t of coal and releases 483t of CO_2 into the atmosphere. Considering that the base of a wind turbine requires at least 200t of cement, the total CO_2 increases to almost 2000t, and this number still does not take into account the CO_2 emitted by transporting these large and heavy components to their installation site. A large wind turbine produces only a maximum of ~ 5MW of electrical energy. Thousands of these turbines are needed and, even then, they cannot satisfy the demand for renewable electrical energy. It is obvious that wind generators cannot prevent the advance of global warming.

Steel and cement production alone contributes 16% of CO_2 to global emissions.

2) OIL

Food production is highly dependent on oil, because the manufacture of synthetic fertilizers requires vast amounts of petroleum derivatives; one ton of fertilizer, on average, generates 800kg of CO_2. In addition, agricultural machinery and transportation and, as previously mentioned, the flatulence of cows and other herbivores (goats, sheep, horses), raise the total CO_2 emitted by all agricultural activities to 30% of global emissions.

Petrochemical products are essential for the production of synthetic fibers, which are basic for most garments, since not enough natural fibers, such as cotton, wool, etc., exist to dress all human beings. The wide variety of plastic products must also be taken into account, such as auto parts, furniture, containers, food packaging, bags, etc. All of these products come from oil, although their production generates little CO_2.

Large-scale commercial shipping mainly uses the cheapest diesel and thus not only emits large amounts of CO_2, but also a lot of sulfur dioxide (SO_2) and nitrous gases (NO_x). These three gases are responsible for the acidification of the oceans. Some of the largest companies, such as Maerks, should be commended for building their new ships to use biomethanol, which produces 30% less CO_2 without producing acidifying gases. Almost all large warships use nuclear propulsion, but due to its greater operational complexity, it is not yet used in commercial transports.

Commercial aviation, for its part, is totally dependent on jet fuel. Large airplanes flying the transoceanic routes in very high regions of the atmosphere deposit their exhausted CO_2 directly in the worst place, where this greenhouse gas is most harmful. Even today, nuclear or electric airplanes are only a dream.

3) SUMMARY

In the most optimistic scenario, all land transport may be electric within a few years, but clean and renewable energy from all possible solar panels and wind turbines can meet only 30% of the world's electricity demand. The deficit must be covered with energy from thermoelectric or nuclear power plants. To ensure the survival of humanity, we have to reduce CO_2 emissions by 50% within the next 30 years. It is of the utmost importance that we attack this threat urgently and develop a new source of abundant and clean electrical energy. To this end, we are left with no alternative but to install new and safer nuclear fission reactors, until nuclear fusion, which replicates the process of the Sun here on Earth, becomes available (next chapters).

VII

CLEAN RENEWABLE ENERGY

---

1) PHOTOVOLTAIC POWER

In 2019, photovoltaic solar energy, connected to the grid, was one of the world's fastest growing sources of energy, reaching a total installed capacity of 630GW (gigawatts = 1000MW). As solar panels have become more affordable, their production has increased exponentially since the beginning of the 21st century, doubling approximately every two years. China has become the world's largest photovoltaic producer, and with 45GW installed, but unfortunately, new thermoelectric plants are also being installed at the same time to meet the country's growing electricity demand. The World Bank works in partnership with the world's governments to strengthen energy sector institutions and develop legal frameworks to improve policies and regulations, helping to lay the groundwork for a major expansion of solar and wind energy in many countries. The International Finance Corporation (IFC) strongly supports renewable energy and its value chain, facilitating financing for equipment manufacturers and companies to lessen their carbon footprint. Several countries, especially in Europe, have set targets to ensure that a certain percentage of the electricity they consume comes from renewable sources.

Solar panels convert the sun's energy into heat or electricity. Infrared light heats water, circulating through a simple cell, for domestic heat use. More complex photovoltaic cells absorb photons from light and convert them into direct current electrical energy that is sent to a converter to generate alternating current. The most commonly-used panels consist of many units with small photovoltaic cells, with the following measurements: 165cm high, 100cm wide and 4cm thick. Capacity and price depend heavily on the technology used, and the energy produced varies between 190-330W per panel. Panels that use monocrystalline silicon semiconductors yield more energy than cheaper options, made with polycrystalline or amorphous silicon. A medium-sized house has an average demand of 3kW and thus requires 10 solar panels. Generating 1MW requires 4256 units and a one-acre (4,047m²) piece of land. The panels are installed not only on the roofs of houses, but also for industrial use. They are found in the large roof areas of factories and warehouses. In cities, houses with panels are generally connected to the electricity grid and thus the excess energy generated during the day can be returned to the grid for a credit. In remote regions without any connection to the electricity grid, the energy generated during daylight hours passes through regulators to banks of multiple batteries, which represent a large investment. Today, solar panels are already used in traffic lights, public lighting and to illuminate many open spaces. Many power companies take advantage of their large unoccupied spaces for large-scale panel installations. In China, the so-called "Great Wall of the Sun" occupies an area of 1,200 km². Manufacturers typically guarantee a 25-year life for their units. Solar panels have an energy efficiency of 15-20%. To achieve optimal performance, the panels are installed facing south and must be positioned correctly based on the latitude of the site. In a few areas, groupings of panels automatically orient themselves to the movement of the Sun.

Each small photovoltaic cell is basically a sandwich made up of two slices of semiconductor material, generally silicon, "doped" with another element to give each piece a different electrical charge (+/-). This generates an electrical field at the junction between the two pieces. When a photon from the Sun hits a free electron, the electrical field pushes this electron out of the silicon, generating an electric current. Metallic plates on the sides of the cells collect the electrons and transfer them to the wires.

The manufacture of solar panels is a long and complicated multistage process. Their production is highly energy intensive because elemental silicon, in amorphous form, is produced from quartz which is silicon dioxide (SiO_2, sand), reduced with coal in an electric arc furnace at 1500°C. The resulting purity of 99% is insufficient for electronic uses.

This metallic silicon is transformed through chemical processes into silicon tetrafluoride (SiF_4), a volatile liquid that, through several distillation stages, is fractionated until the impurity concentration is less than 0.2ppm, which is semiconductor grade silicon. This liquid is transformed into metallic silicon in a reaction with hydrogen at 1100°C and then melted. By lowering the temperature very gradually, an ingot-shaped single crystal is formed. A more accelerated process produces polycrystalline silicon of lower efficiency. The next stage, cutting the ingots into wafers with a thickness of 2-4mm (millimeters), results in losses of 50%. This is followed by multiple steps - texturing, "doping," assembly, etc. - until an individual cell with a surface area of 7.5cm² is completed. With sufficient lighting, each cell is capable of producing a differential of 0.4V and 1W of electrical energy. Multiples of these photovoltaic cells are mounted in a frame to form a panel. Semiconductor grade silicon is the basis for transistors used in all electronic equipment. The manufacture of solar panels uses coal and requires a lot of electrical energy, much of it still produced from fossil fuels. The process has a harmful effect on the environment. According to estimates, a solar panel has to generate clean electricity for twelve years to compensate for the dirty energy used and CO_2 emission in its manufacture.

Photovoltaic technology is still a relatively new invention. Many companies are dedicating significant human and financial resources to improving energy efficiency and reducing energy losses. To find which areas have the potential for improvement, we must first have a good understanding of the physics of the photoelectronic process.

- A new photocell uses 20% of the light it receives but after only few hours of operation it loses 10% of its efficiency, an effect that is not yet understood.
- The light received by the very thin upper silicon slice can be reflected, absorbed or pass directly through the transparent piece. To absorb the maximum amount of incoming light and minimize reflection, the surface is covered with a mono-molecular layer of silicon dioxide or titanium dioxide through a process called vapor transfusion.

- Each of the two silicon slices is treated separately with a "dopant." Phosphorus (P) is used in the upper layer for a positive charge and boron (B) in the lower layer, resulting in a negative charge. These new elements are incorporated into the crystal lattice of silicon atoms.
- Semiconductors are "semi-metallic" elements that can exist in two forms, one metallic and the other non-metallic. They belong to a group of specific elements in the periodic table of elements that includes silicon, germanium, tin and others. In its non-metallic form, the element silicon can form transparent crystals. A special property of these elements is that, due to their atomic structure, the electrons in their superior orbit can be released when hit by light (photons) of a specific frequency.
- The light that hits the silicon atom releases an electron from the upper orbit of its electrons, which moves freely between the atoms. Doping elements of different polarities (P + / B-) form an electromagnetic field that accelerates the free electrons and creates a voltage of 0.4v and 1W of power. Therefore, 36 photocells are needed to produce an open circuit of 20v.
- The process of doping a monocrystalline silicon ingot is very complex, because a perfect homogeneous distribution of the doping agent is required. This can only be achieved through neutron irradiation with a special device. World demand for doped silicon at the moment is 200 tons per year and growing very fast.
- The upper crystalline lattice of silicon atoms includes a phosphorus atom with five electrons in its orbit, which results in one electron left over in the doped lattice, thus producing a negative charge. In another doping phase, a boron atom is added into the lower slice of the silicon lattice. The addition of boron, with three electrons in its orbit, means that this crystal lacks an electron, giving this slice of silicon a positive charge. The difference in charges produces a voltage of 0.4 volts.
- Other more complex semiconductors also exist, such as gallium arsenide or silicon alloys with germanium or gallium, which provide greater electrical efficiency, but their high costs limit their use mainly to space facilities, such as satellites. Silicon, as the most

abundant element on Earth (in the form of sand) after oxygen, is the material of choice in all electronic applications, especially for high-speed integrated circuits.

– A photocell's energy efficiency depends on several other factors, such as the light's temperature and wavelength, with higher efficiency at lower temperatures. Therefore, in a few installations, especially on roofs, water is allowed flow over the cells. Obviously, this is impossible in installations in desert areas. Silicon photocells respond only to a specific frequency of the light wave. They do not absorb high-frequency light, such as ultraviolet light from the Sun or low-frequency light, such as infrared light that only produces heat. The most expensive and energy efficient photocells have a thin film that absorbs ultraviolet light and another that captures infrared light and thus converts the entire spectrum of visible light into electrical energy. Naturally, it is important to keep the surface of the solar panels clean and free of dust, etc.

– Normal solar panels (160x100cm), made with monocrystalline silicon, produce between 150-200W of energy. Smaller panels, using the cheapest polycrystalline silicon, are used for solar calculators or garden lamps, etc.

The eight points listed above represent just one example of the many opportunities for improving the electrical performance of solar panels.

2) WIND POWER

Wind energy has become a key source of electricity generation for changing to a cleaner and more sustainable energy model. Improved technology enables some wind farms to produce electricity as cheaply as coal or nuclear power plants. The simplest way to explain wind energy is that a wind turbine works just the opposite of a fan. Whereas a fan uses electricity to make wind, the wind turbine takes advantage of the wind to make electricity.

Wind power uses the force of the wind to produce electricity through a wind turbine that drives an electric generator. The kinetic energy of the wind drives the three oblique blades, attached to a common axis of the wind turbine, and converts it into rotating mechanical energy. The gearbox, through several stages of multipliers, increases the speed of the rotation of the shaft of a three-phase alternator to 1000-2000rpm (revolutions per minute) that produces alternating electrical energy. Normally, for a wind turbine installation to be profitable, at least twenty must be grouped in one-hectare wind farms, sharing the necessary regulators, transformers, etc. to feed the energy generated (690v) to the electrical grid from 20 to 132kv (kilovolts). To avoid interference from air turbulence after passing through the rotor, wind turbines are installed at a distance of 3-10 times the length of their propellers. Still, wind farms require less land per unit of electricity produced than fields with solar panels.

Winds are generated by non-uniform heating of the earth's surface; 2% of the energy from the Sun is converted into wind. To ensure good performance from the wind turbines, it is essential to carry out a study prior to the placement of the wind farm. It is important to know the diurnal and nocturnal variations, the direction and seasonal changes of the winds, as well as the air speed at different heights, because the roughness and friction with the surface and obstacles such as trees, buildings, etc. reduce wind speed and produce turbulence. Therefore, the tallest generators are the most energy efficient: energy production increases 0.5% for each meter of height. Offshore wind turbines take better advantage of the uniformity and force of the wind over the sea and thus are very common in northern european countries. Due to the inertia and friction of the multiple moving parts inside the wind turbine, in order for the blades to move, the minimum air speed needs to reach 14 km/h. At 90 km/h, a limiter in the internal gearbox stops the rotation of the propellers to prevent damage from excessive centrifugal forces. Wind speed determines the number of revolutions per minute (rpm) of the rotor, typically between 8-30rpm.

The towers of many early wind turbines, still in service, have a height of between 50-80 meters. More recently, taller wind turbines, up to 150m high, have been built. With their long propellers (90m), these installations reach a total height of 250m and the rotor exposes a 25,000m² (πr^2) circle to the wind. The cost/benefit ratio is improved with the larger sweep area of the turbine, which is the circle covered by the propeller. The speed at the end of the longest blades, turning at 30rpm, is very high, up to

300km/h. This is the place where the blades wear out due to the high air flow, which ultimately limits the life of the equipment to twenty years. To avoid unbalanced rotation, the face of the rotor must always be directly oriented to the wind. In smaller equipment, this is ensured using a vane installed above the nacelle. Larger turbines use a direction sensor and are guided by servomotors.

The rotating blades of the tallest towers measure up to 90 meters long with a weight of 50+ tons each. Due to their complex oblique geometry, their manufacture in a single piece requires special equipment and materials such as polyester or epoxy reinforced with fiberglass and carbon fiber for limited flexibility. A wooden structure inside ensures the proper rigidity. Just transporting these huge parts to their installation site costs 20% of the total cost of the wind turbine. In the newer wind turbines, the blades are also rotated to take full advantage of wind energy. A severe ice storm, like the one that recently covered most of the United States, can cover wind turbines with ice and shut them down for a few days.

Most of the wind turbines still in use have a capacity of 3-5MW. The larger, more modern turbines can produce up to a nominal 10MW when a brisk wind is blowing. The average cost of a modern wind turbine is US $ 45 million. A 10MW wind turbine can meet demand for 9,000 homes. The cost of an offshore installation is normally 30% higher. Wind farms, due to the variability and intermittency of the wind, always require a reliable backup of a quick-start thermoelectric or hydroelectric plant to guarantee a continuous and reliable supply of energy.

After hydroelectric energy, wind is the energy that produces the least greenhouse gases. Its manufacture, construction, transportation, etc. generates only 12 grams of CO_2 for each kWh (kilowatt-hour). Despite the great environmental advantages, certain problems also come with wind energy, such as noise from the aerodynamic whirring of the rotor. Wind turbines also cause the deaths of many birds, especially nocturnal raptors, and bats.

In Mexico, the CFE (Federal Electricity Commission) has exclusive control of wind energy, installation, production, etc. The country has 8 wind farms in operation and many others under development. ~ 5% of the country's electricity demand is met by wind powe

3) FISSION ENERGY

Nuclear power plants that use uranium or other radioactive heavy elements can be considered to generate renewable electrical energy, because they do not produce any greenhouse gases and their raw material exists in abundance. Uranium is 100 times more common on Earth than silver, and the mineral is widely available in large deposits on every continent on Earth. Thus, uranium can be considered a suitable source of renewable energy.

Due to the exponential growth of the global population, energy requirements cannot be satisfied only with clean renewable energy, such as wind turbines and solar panels, especially since these resources are intermittent and require a backup to guarantee continuous and reliable electricity. To meet this deficit, new thermoelectric plants are still being built today. These plants use natural gas in places where pipelines exist, but others continue to rely on fuel oil or, worse still, coal. All of them produce more greenhouse gases, worsening climate change. The only alternative is new nuclear fission plant designs, which must be inherently safer, especially since they eliminate the element of human error. Nevertheless, the danger of radioactivity persists, along with highly radioactive waste. Modular construction can reduce the high cost and long construction time; however, we must also overcome the great public resistance resulting from serious accidents in the past. Until nuclear FUSION is commercially available, in maybe twenty to thirty years, nuclear fission reactors are the only intermediate solution to save the Earth from global warming.

Natural uranium is mainly a mixture of two isotopes, 99.3% U-238 and 0.7% U-235. Uranium ore, as found in nature, is slightly radioactive; all the various isotopes have a half-life of millions of years. In their slow radioactive decay, they emit alpha and beta particles, plus electromagnetic radiation, such as X-rays and gamma rays. When absorbed into the earth, these rays produce heat. This energy helps to maintain the high internal temperature of the Earth.

The word "nuclear" immediately recalls the catastrophic disasters of Chernobyl and Fukushima. For fear of more accidents, the construction of new nuclear power plants has been slowing down around the world. Germany decreed that, in 2022, all 17 of its nuclear power plants, which generated 22.4% of its electricity demand, must be deactivated. Accelerated projects to generate renewable energy could not satisfy national demand; as a result, Germany had to build another thermoelectric plant using natural gas, releasing more CO_2 into the atmosphere. In France, in contrast, nuclear energy meets 76% of national electricity demand. The plants in France are close to large urban centers and have never had an accident. Of the six nuclear accidents in the world which emitted radioactive gases, the two worst were caused by poor design combined with serious human error. They resulted in intense international debate about the feasibility of nuclear energy.

To overcome this resistance from the population and to aid a logical discussion of nuclear power, it is important to understand the root causes of these two catastrophes. First, absolutely, they were not nuclear explosions. A nuclear explosion can only occur with U-235 enriched to more than 95%; in nuclear reactors, U-235 is enriched to a maximum of 5%. The two nuclear power plants in question used boiling water system reactors, not pressurized reactors, in which water circulating through the reactor core carried the heat from the core in the form of steam directly to the turbines to generate electricity. In both cases, the fundamental cause was the failure of water recirculation through the core. In the Chernobyl accident, this was due to serious human errors; in Fukushima, the reason was a total electrical failure resulting from the tsunami. Without cooling, the cores reached temperatures well above 1,500°C, and the remaining water thermally decomposed into free hydrogen, which caused detonation and opened the equipment to the air. The molten core emitted the highly radioactive elements of the nuclear reaction into the atmosphere. Today, with the third and fourth generation of reactors, radically new designs have been implemented that guarantee absolute equipment safety, but the human fear factor is still an unpredictable element.

Nuclear reactors use U-235 enriched to 5% through various physical separation processes. Atomic bombs use highly concentrated U-235 at > 95%. U-235, with 143 neutrons and 92 protons, is the natural element with the highest atomic weight. Its oversized nucleus gradually decomposes (fission) into many smaller, highly radioactive elements, releasing heat and radiation. The total mass of the products of this fission is slightly less than the mass of the uranium used. This small difference in mass is converted into a lot of energy, according to Einstein's famous equation, e = mc²: the deficit in mass (m) is converted into energy (e) multiplied by the square of the speed of light (c²), which is 300,000km/second (factor of 90,000,000,000). One gram of U-235 produces 10 million times more heat than burning a gram of coal.

When U-235 absorbs a neutron, the atom disintegrates into two lighter elements, such as cesium-135, strontium-90 and iodine-129, which are highly radioactive. This fission emits 2-3 neutrons at very high speed. In the reactor core, the fuel consists of only 3-5% U-235; the rest is U-238, which is not fissionable. This U-238 captures the high-speed neutrons and becomes plutonium, PU-239. Fission cannot therefore be maintained, because U-235 does not capture high-speed neutrons. For this reason, the core of a commercial reactor contains a moderator, such as water or graphite, that slows the speed of these neutrons and thus can fission 2-3 more atoms of U-235. With the proper concentration of 3-5% U-235 in the fuel, this fission turns into a controlled chain reaction that produces a lot of heat.

A fission reactor consists of the following essential parts:

3-5% U-235 fuel (the rest is U-238); the usual fuel in water-cooled plants is uranium dioxide (UO_2) in the form of small cylinders.

Fuel elements are hermetically sealed zirconium rods containing the UO_2, these rods are grouped in a rectangular arrangement of 8x8 units.

Neutron moderator: Water or graphite, which functions to slow down the speed of the neutrons produced by fission so that they can interact with other fissile atoms, U-235, and maintain a controlled chain reaction.

Cooling: a recirculation system, usually water, that conducts the heat generated to a heat exchanger, or directly to the electricity generating turbine.

Reflector: water or graphite that reduces the escape of neutrons and thus increases reactor efficiency.

Control rods: stainless steel tubes containing compacted cadmium or boron carbide; they are very efficient neutron absorbers and thus can control the speed of fission or stop it in an emergency.

In modern plants, safety elements consist of multiple active systems that respond to electrical and passive signals which act naturally (gravity). The cylindrical concrete structure surrounding the reactors, heavily reinforced with steel and topped with a dome, is the ultimate safety precaution to prevent a release of radioactivity into the atmosphere.

The different configurations of the reactors in nuclear power plants contain a load of 30 tons of fuel that must be changed every 5 years, leaving the same amount of highly radioactive waste. This waste still produces a lot of heat and, therefore, needs to be stored in large amounts of water in refrigerated pools for 5 to 10 years until cool enough for transfer to large cement containers. This waste remains extremely radioactive for thousands of years and has to be stored in very deep geological locations that are safe from earthquakes. Such storage is very expensive. In 1945, a 66-cubic-meter deposit of highly radioactive waste exploded in Chelyabinsk in the Ural Mountains, severely polluting a 100-hectare area, reducing 23 villages to rubble and forcing the evacuation of 10,000 people.

The world has a total of 454 nuclear reactors, with a capacity of ~ 400GW. They are in operation in 31 countries and satisfy 16% of the world's electricity demand. More than 53 units are under construction, of which 16 are in China. 115 have already been withdrawn from service due to technical problems or because they have reached the end of their useful life, which is 25 to 40 years. Several other plants were never completed. Of the 115 deactivated reactors, only 19 were completely disassembled. The others were abandoned in situ and left to cool down very gradually over decades. Due to the dangers of high radioactivity, disassembly is more expensive than the construction of a new plant.

Mexico has two nuclear reactors at Laguna Verde in the state of Veracruz with a capacity of 1,604MW, which should satisfy 3.2% of the country's electricity demand. Unfortunately, this plant is plagued by frequent minor radioactive leaks, technical failures, improper waste management, and a continuing lack of resources to correct these problems.

In general, the capacity of a nuclear reactor is 1,000-1,500MW (1-1.5GW, gigawats. 1GW = 1000MW). Normally four of these units are grouped in a nuclear power plant generating between 5 and 8GW. This capacity can replace 10 to 16 standard thermoelectric plants. Thus, a nuclear power plant avoids the emission of 10.4-22.4 million tons of CO_2 per year into the atmosphere. The cost of a 1GW nuclear reactor is US $ 6-9 million. Construction alone takes 4-6 years, but environmental studies and permits can cause long delays and high additional costs. Nuclear reactors produce energy constantly and thus are a perfect backup for the intermittency of solar panel and wind turbine installations.

Safe nuclear reactors have been developed in all the necessary sizes, mainly for propulsion. Most warships use them. To avoid the large amounts of CO_2 emitted by merchant shipping, at least large carriers should use these modular reactors. Satellites and space probes use solar panels or nuclear reactors, especially small ones, to generate their electrical energy. Nuclear rocket experiments remain unsuccessful.

The existing nuclear power plants now in operation are the only option to satisfy the deficit in the demand for renewable energy, despite the operational and cost difficulties. Nuclear FUSION (next chapters), which promises to produce clean and abundant electrical power, requires at least an additional 30-50 years before its successful commercialization. Thermoelectric plants take a year to build and cost a fraction of nuclear power plants. For this reason, some countries are still building new thermoelectric plants that continue to worsen global warming. Recently, the president of Mexico "proudly" announced the construction of a new thermoelectric plant using fuel oil in Baja California.

— New developments

In the nuclear industry, scientific knowledge and technology are continually advancing, causing safety requirements and standards to evolve based on new knowledge and acquired operational experience. The nuclear industry is one of the most technologically advanced industrial sectors, comparable to the aeronautical and space industry. The urgent race to find new safe sources for generating clean and reliable electrical energy helped develop the third and fourth generation nuclear reactors.

Generation III plants, designed through the development of existing generation II plants, bring together evolutionary improvements based on experience acquired in the operation of current reactors. These advances include, above all, passive safety systems that act on natural convection or gravity and are designed to deploy by themselves when operations deviate from their normal mode, without anyone or anything having to activate them. And they work without electricity. All new constructions must use this technology to get permits.

Generation IV nuclear power plants, quite large prototypes with a capacity of up to 350MW, are in the demonstration phase. New designs should be ready for commercial construction in 2030. The most promising system is the Fast Breeder Reactor, cooled by metallic sodium. At the Beloyarsk nuclear power plant in Russia, Units 3 and 4, each with a capacity of 200MW, use this design. A similar, larger reactor is also in operation in China.

The generation I and II nuclear reactors use U-235 enriched to 5% as fuel. The rest is non-fissionable U-238. Only part of the U-235 is burned, resulting in only 2% of the uranium's potential energy being converted into useful heat. In fast breeder reactors, the fuel arrangement allows U-238 to absorb fast neutrons and converts it into the fissionable plutonium, Pu-239, which is directly involved in fission, producing much more energy. These reactors produce more plutonium than they consume, hence the name "breeder." They are capable of using 60 to 70% of the uranium. Given the high operating temperatures, a large amount of heat needs to be removed. For this reason, liquid sodium is used as a refrigerant, passing through two heat exchangers where water vapor is produced for the electric generators.

The energy production in the core of a fast breeder is intense and therefore the coolant must have very good heat transfer characteristics. Metallic sodium, compared to other possible refrigerants, presents the best combination of the properties required, such as low pumping power, operation at atmospheric pressure, and very good neutronic conditions. This type of reactor burns most of its fuel, minimizes waste and has a very long life of up to 60 years.

Fast breeder reactors operate at temperatures up to 600°C. This higher energy density allows for a more compact design and greater modularity: Large system components can be built in other factories and moved to the installation site, reducing construction times and costs. One of the novelties of generation IV reactors is that the designs do not have to be exclusively oriented to electricity generation. Due to their high operating temperature, they could produce hydrogen fuel for transport systems.

After the successful demonstration of the Russian prototypes, new sodium-cooled fast breeder reactors were installed in the United States, France, Japan and Germany, with steadily increased capacity up to

600MW. The following factors favor the construction of this type of power plant:

- Much less nuclear waste that only lasts a few centuries instead of millennia.
- Higher energy efficiency (> 100x) and a high degree of burning.
- Possibility of consuming existing nuclear waste.
- Greater operating safety, simplicity of operation and stable controls.

Some sodium leakage problems and a related fire in Japan are delaying further commercialization of these types of reactors. Meanwhile, Bill Gates is investing millions of his own money into a project called TerraPower to develop an innovative reactor technology, transforming U-238 into fissile plutonium, known as a traveling wave reactor. This reactor uses molten salt to carry the high heat produced through heat exchangers to electric generators. Initial results are encouraging but have yet to be demonstrated in a reliable prototype.

VIII

NUCLEAR FUSION ENERGY - A TINY SUN IN A MAGNETIC BOTTLE

---

1) WHAT IS NUCLEAR FUSION?

Generating energy from the Sun, directly here on Earth, does not imply using solar panels or wind turbines, but rather involves creating a small sun in a closed device to generate abundant electrical energy without any CO_2 emissions or radioactive waste. In Greek mythology, Prometheus stole fire from the gods and brought it to Earth. Today, we are working to reproduce on Earth the process that our Sun uses to generate its intense energy. Nuclear fusion is the magnificent process used by the Sun and most of the stars in the Universe, to transforms hydrogen into helium, an inert gas, with the release of an enormous amount of energy. This energy was already demonstrated here on Earth in 1952 with the hydrogen bomb, unfortunately in a very destructive manner.

Nuclear hydrogen fusion begins in the inner region of the Sun, where, due to its great mass (330 times the mass of the Earth), the pressure is an enormous 10^{11} atmospheres. The gas in the Sun exists in the form of plasma at 15^0 million C (centigrade), where hydrogen atoms lose their electrons and exist as protons with a positive charge: Plasma. Protons, with a positive charge, repel each other through the electrostatic Coulomb force, but the high temperature accelerates them until they collide and the high pressure compresses them until they are close enough together, at which point the nuclear force fuses them. This force, which is 100 times stronger than electrostatic repulsion, acts when two positively charged particles come close enough together. Four hydrogen protons fuse to form a helium atom, whose mass is slightly less than the mass of the four initial hydrogen atoms. This small deficit of mass is converted into a lot of energy, according to Einstein's formula, which explains that mass is converted to energy with a factor of $300,000^2$ (c^2). Heat, or thermal energy, is also considered to be the kinetic energy, or velocity, of the particles.

For about 80 years, many small and medium-sized companies and several universities have been working with prototypes to develop a commercial fusion reactor. It has almost become a joke to say that commercial fusion always seems to be about thirty years away. Unfortunately, all these projects are working concurrently without any open exchange of scientific information or any technical coordination, and the smaller ones have been limited by a lack of money. In recent years, the threat of climate change has resumed and strengthened interest in nuclear fusion, because all other projects for generating clean, renewable energy cannot meet the growing global demand for electricity. Even the new nuclear fission reactors do not offer an ideal long-term solution, due to the accumulation of up to 400,000 tons of highly radioactive waste that must be stored for thousands of years and for which a safe place has not been found.

To give new impetus to nuclear fusion, the "Commonwealth Fusion Systems" (CFS), led by prestigious researchers from MIT (Massachusetts Institute of Technology) was recently established with the ambitious goal of developing a commercial fusion reactor within fifteen years. This company is financed by a large investment fund, in which Bill Gates, Jeff Bezos, Jack Ma (the Chinese owner of Alibaba), Richard Branson and many other well-known international billionaires have holdings. Stephen Hawkin also helped with this project. It is investing in various projects in all developed countries working on nuclear fusion. As of today, it has already invested about 80 billion dollars. Together with Google, the company has created a computational algorithm to optimize its experiments, guaranteeing a free exchange of new ideas and a measure of coordination. A dozen startups are already competing to be the first to build a nuclear fusion reactor. The most ambitious project along these lines is ITER, which involves 35 countries, including the United States, Russia, China, Japan, South Korea and the European Union. The project aims to build "the largest Tokamak in the world" in Caderache in the South of France. After several delays, the first plasma tests are scheduled to begin in 2035. So far, $22 billion has been invested in this project.

To facilitate the fusion process here on Earth, most of these processes work with deuterium (D) plus tritium (T). Both are isotopes of hydrogen ($_1^0H$), which has only one proton with one electron in its orbit, no neutron (n^0). Deuterium is a hydrogen atom with one proton but with one additional neutron ($_1^1D$). Tritium is a hydrogen atom that has one proton and two neutrons ($_1^2T$). Deuterium is the heavy water molecule (D_2O), widely available in seawater and used as a moderator in various nuclear fission reactors, where tritium is also produced. This atom is radioactive and has a half-life of 12.5 years.

The very high pressure inside the Sun cannot be reached here on Earth. To compensate, much higher temperatures of 150 million °C must be used with atoms that can be fused more easily, such as deuterium and tritium. Using higher temperatures imparts more kinetic energy to the atoms and allows nuclei to approach each other over shorter distances, where the force of nuclear attraction exceeds the forces of electrostatic repulsion. Under this condition, a deuterium combines with a tritium, forming a helium ($_2^2He$) and releasing a high-energy neutron. This neutron reacts with a deuterium, regenerating a tritium and thus the fusion continues, only feeding more deuterium into the plasma.

Fusion of deuterium with tritium:

$$_1^1D + _1^2T \text{ ---- } _2^2He + _0^1n$$

Tritium regeneration

$$_0^1n + _1^1D \text{ ---- } _1^2T$$

The high temperature produces a gaseous mass called plasma, made up of free electrons and ionized atoms. The plasma must be confined and controlled at high temperatures in the cavity of a fusion reactor (Tokamak) for the time necessary for the reaction to take place. Two methods of confinement have been developed:

- Inertial Confinement Fusion (ICF): consists of creating a medium so dense that the particles have to collide with each other. A small sphere of deuterium and tritium is impacted by groups of powerful lasers, whose simultaneous discharges cause an implosion. The ball of these atoms becomes hundreds of times denser, allowing fusion to take place. Several reactors of this type are currently in development.

- Magnetic Confinement Fusion: the plasma's electrically charged particles are trapped and controlled in a reduced space by strong magnetic fields, to achieve a sufficient density of the plasma to force the nuclei close together until they fuse. The most advanced device for this reaction is the Tokamak.

The Tokamak is one of the preferred models for fusion. It is a toroid (doughnut)-shaped vacuum chamber that contains deuterium. It is heated with strong electrical discharges to very high temperatures of 100-150 million C^0, where the gas ionizes (loses its electrons) and reaches the plasma state. A strong magnetic field, with powerful superconducting electromagnets, confines and compresses the plasma for fusion to begin. This reaction has no danger of explosion, because when the external electrical energy is turned off, the fusion reaction ends. The fusion product is a little helium gas, which is inert and has no greenhouse gas effects but many industrial uses. The function of a commercial fusion reactor is for its heat to generate steam that drives turbines to produce electricity. As of today, the energy balance is still below one (<1), which means that the energy produced is less than the power used. For commercial use, this balance must be at least double (> 2).

Inside the torus of the Tokamak, the particles circulate with high kinetic energy in the form of a ring at temperatures of 150 million ^{0}C. This hot plasma is controlled by two strong, vertically and horizontally oriented magnetic fields. Tokamaks are equipped with two multiple electromagnet arrays that produce these strong magnetic fields because they are cryogenic superconductors, cooled by liquid helium at a temperature of -267.9^{0}C. (5.2^{0}K). At this temperature, electrical energy circulates through the electromagnet wires without resistance and without overheating them. This controls the stability of the plasma and prevents hot particles from escaping and staying away from the metal walls. Most of the experimental Tokamaks have walls of tungsten (with a melting point 3,422° C) that heats up to only a few hundred centigrade. In a Tokamak, having good control over the flow of heat is critical to bring the high temperature generated by the fusion from the equipment wall to a practical use in producing electricity. The temperature profile of a Tokamak is extreme: 150 million C^0 in the plasma. A short distance away, at the wall, the temperature is no more than 600^0C. Outside the wall, the temperature of the electromagnets is -267.9^0 C.

The objective of the gigantic ITER project (International Thermonuclear Reactor) is solely to function as a scientific experiment, demonstrating the technological feasibility of nuclear fusion so that, later, the participating countries can use this new technology in their own power plants. To build the giant Tokamak for this experiment, different countries have worked together on the multiple, huge and heavy components based on their specialties, sending them in special ships to the port of Marseille. From there, they are transported on a special 100-kilometer highway to the installation site. The logistics coordination behind the construction, delivery, installation and, especially, international financing of the project is complex, and problems have already arisen related to timing and a lack of human resources. Among the 36 participating countries, Europe is paying 45%, with the others each bearing 9% of the costs. As of the year 2020, a total of US $ 22 billion has been spent. The total cost is estimated at 65 billion. Despite multiple delays, the current estimate is for the first plasma to be turned on in 2035. The fusion must use 50MW to produce 500MW, a production factor of 10. It is hoped that experiments with smaller Tokamaks can help resolve the various pending scientific questions to be incorporated into the large project and thus achieve a successful conclusion.

Meanwhile, nuclear fusion is getting more and more public attention. Even some large oil companies are participating in projects with medium-sized Tokamaks (~ 8x3m), which have already resulted in many new scientifically important results. A new reactor, called the "Stellarator," was built recently in Germany with the assistance of two supercomputers and has ignited its first plasma. In this reactor, a special geometric arrangement of vertical superconducting magnets introduces a continuous torsion into the plasma, further compressing the ionized gas ring and concentrating the fast-moving neutrons from the fusion to impact and react with the tritium instead of escaping. Similar results were obtained from the so-called "Jet" project at Oxford in England, indicating that more compact, and consequently cheaper, reactors may be preferable. The escape of neutrons at high speeds from the fusion can cause a problem for ITER, because neutrons that hit the metal wall are incorporated into the atomic structure of the metal and make it brittle and radioactive.

2) THE URGENCY OF DEVELOPING NUCLEAR FUSION AND HOW TO DO IT

To avoid catastrophic climate change due to global warming, clean electrical energy must be generated as quickly as possible without burning fossil fuels, limiting CO_2 emissions to a basic level that our environment can manage through biological processes, such as photosynthesis, etc. According to the most optimistic forecasts, it will take 30 to 50 more years for a commercial fusion reactor to be built, but climate change only allows us 10 to 20 years to control greenhouse gas emissions before global warming will become irreversible.

As has already been shown in previous articles, clean and renewable energy from non-polluting facilities is insufficient to satisfy the world's electricity demand. Today, the only temporary solution is to build more nuclear fission reactors. It is true that these do not produce CO_2, but, as explained before, they do come with certain disadvantages, such as radioactivity hazards, potential for contamination by radioactive waste, high costs and multiple years of construction, etc. For this reason, it is urgent that we implement the necessary actions to coordinate the many different nuclear fusion experiments, facilitating a free global exchange of scientific information and focusing all scientific and technical efforts on one or two multinational projects without any limitation of funds.

In view of this urgency, the following resolutions should be taken:

During World War II, the United States was in dire straits when Einstein tipped off President Roosevelt that Germany was working on developing an atomic bomb. Under this threat, the ultra-secret "Manhattan" project was founded. With unlimited resources, 300,000 scientists, engineers and technicians worked together and within three years, they exploded the first atomic bomb. Obviously, impending danger can force a nation to achieve the impossible and overcome all obstacles. This success should motivate the entire world to concentrate all its efforts on developing nuclear fusion to limit the increase of temperatures on Earth.

How can a similar effort be organized to commercialize a nuclear fusion reactor in less than 20 years instead of the expected 50 years? This would save humanity from catastrophic calamities resulting from the continuous increase of CO_2 from the generation of electricity, which will cause Earth's temperature to increase by another 1.5° C and make climate change irreversible.

Financial Resources: Several large companies in the United States are spending billions of dollars to bring human beings to the planet Mars and establish a colony there within a few years. In this case, the big question should be: Wouldn't it be more useful to dedicate these human and financial resources to save our Earth first? Also, in different countries, too much money is spent on detecting life on planets outside our solar system. In view of the threat of climate change, it is much more important to preserve life on our planet.

Human Resources: At the large Hadron Collider in Cern, Switzerland, some 3,000 eminent scientists are working on proton collisions to determine their subatomic components. This experience could go a long way towards solving the last questions still remaining in facilitating particle collisions and fusion in a stable plasma. Involving these experts in the scientific problems of nuclear fusion could considerably advance the success of nuclear fusion.

International Resources: It is evident that a single nation, in a short time, cannot build all the multiple components necessary for a commercial-size Tokamak. Different countries have demonstrated their specialties: some in building large metal structures, others in developing cryogenic electromagnets, others in manufacturing complex electronic control circuits, etc. The logistical difficulty requires a general coordinator; a person respected worldwide, assisted by specialists, to collect and combine all the scientific information; and another person to procure the financial resources. These three individuals, along with their work teams, must demonstrate an almost super human talent to bring a project of this size to fruition in the shortest possible time.

3) ADVANTAGES OF NUCLEAR FUSION

The advantages of a fusion reactor are manifold: the reaction can be stopped instantly and thus is intrinsically safe, there is no danger of harmful emissions, and no radioactive waste is produced. Such a reactor can be expected to produce inexhaustible electrical energy, and this energy will be available at all times. These types of facilities do not have to be located far from urban centers. With sufficient cheap electrical energy, seawater could be converted into abundant amounts of potable water, enough to irrigate desert areas. By generating electricity through nuclear fusion, it will be possible to eliminate all thermoelectric installations, stop dangerous fission nuclear plants, and avoid the need to build more wind turbines or solar panels that produce only intermittent energy. In comparison, nuclear fusion power produces higher energy density, because it requires much less land.

With enough clean electrical energy, it will be possible to convert all land transportation to electric motors, reserving the use of petroleum products for air transportation. These measures should completely eliminate the toxic smog that so seriously affects many urban centers on the planet and harms the health of the population. It will gradually be possible to eliminate the use of natural gas and LP gas to heat residences and businesses and install electrical devices instead.

Despite the growth in renewable energy installations, greenhouse gas emissions are going to increase over the next 20 years, with more and more serious annoyances, not only for humans, but also threatening animals with extinction and agriculture with a loss of biodiversity. Nuclear fusion power plants should stop the generation of more greenhouse gases and gradually reduce the CO_2 accumulated in the Earth's atmosphere, thus preventing the fifth great extinction of life on our planet.

IX

PHOTOSYNTHESIS

Photosynthesis is the wondrous process whereby the leaves of plants take in carbon dioxide and synthesize all the necessary products for living beings through a very complex process. In multiple stages, this process converts inorganic matter, carbon dioxide, water and minerals into organic matter using light energy. The surfaces of the leaves are oriented towards the light and underneath, small pores selectively absorb carbon dioxide, even though its concentration in the air is only ~ 400ppm. Given light energy, the natural green catalyst, chlorophyll, and water, the leaf cells are able to reduce carbon dioxide and incorporate it into the process of producing organic material. The roots of trees seek water and minerals, and the trunk and stems carry it all to the leaves.

The wonder of photosynthesis is that it takes place at ambient conditions as a simple chemical reduction of carbon dioxide. In the chemical industry, the reaction of carbon dioxide is a very difficult process that requires high temperatures and, therefore, a lot of energy. The chemical bond between carbon and oxygen is extremely strong ($O = C = O$), because burning coal produces a lot of energy. Consequently, breaking this bond requires even more energy. During photosynthesis, the plant divides water into hydrogen and oxygen. Hydrogen is used to reduce carbon dioxide and react with water to produce carbohydrates (glucose) that run throughout the plant, providing it with the energy necessary to grow.

Despite the loss of large forests, natural photosynthesis still traps ~ 100 billion tons of carbon dioxide annually. It should therefore be the ideal process for controlling the concentration of carbon dioxide in the atmosphere. Emissions of this gas, however, are far outpacing the natural process and are causing global warming. Given the slow pace of photosynthesis, an industrial-based process must be developed that can absorb large amounts of carbon dioxide to return the atmosphere to a tolerable level (300ppm).

Like land vegetation, phytoplankton in the ocean and algae, especially cyanobacteria, known as "blue-green algae," are capable of photosynthesis. Cyanobacteria float on the surface of water. Chlorophyll absorbs photons from light and releases hydrogen from water to produce oxygen. With free hydrogen, the cyanobacteria reduce carbon dioxide and also trap nitrogen from the air and reduce it to ammonia, which all cells can use. For millions of years, cyanobacteria have been responsible for the evolution of the biosphere and for the oxygen (21%) in today's atmosphere.

The two processes of photosynthesis, terrestrial and aquatic, have been widely studied and well understood for decades. In recent years, it has been possible to genetically manipulate various biological processes to create significant improvements in factors such as yield, quality, resistance to pests, etc. The various biological stages of photosynthesis must be studied to define which ones can be genetically modified to accelerate the process.

It is well known that in the Devonian geological era (~ 416 million years ago), prior to the Carboniferous era, high levels of carbon dioxide caused explosive growth of plants and animals, which then became extinct and formed the coal and oil that we are overusing today. Obviously, photosynthesis can be accelerated with higher concentrations of carbon dioxide. Other parameters that can be investigated to improve the photosynthesis process are: a) chemically changing the catalyst, chlorophyll, to absorb more light energy; b) focusing on the part of the light spectrum that is the most efficient; c) identifying the best environment for photosynthesis (algae seem to grow faster in water than plants do on land); d) determining which species of terrestrial vegetation lend themselves to genetic improvement to absorb CO_2 faster, etc.

X

POTABLE WATER

Climate change has already drastically reduced the availability of potable water on Earth. More than 40% of the world's population has difficulty accessing drinking water. To avoid water wars, we urgently need to develop practical, economical and environmentally acceptable processes to convert seawater (~3% sal) into potable water. Two thirds of the surface of our planet is covered with water. Only 2.5% of that water is freshwater, and only 0.3% is suitable for human consumption.

Two processes can be used to desalinate seawater: reverse osmosis and distillation. Both consume a lot of energy and thus produce large amounts of carbon dioxide. More research is necessary into how biological processes could produce drinking water for the 60% of the world's population that today depends on reverse osmosis, without contributing to global warming.

Reverse osmosis is the most advanced desalination system; 150 countries have plants and 300 million people depend on them. In reverse osmosis, filtered sea water passes at very high pressure (~ 7 bar) through semi-permeable plastic tubes that only allow water (99.5%) to pass through, forcing out a concentrated salt solution. The cost is ~ US $ 1 per cubic meter, which is far too expensive for irrigation. High pressure requires large pumps that consume a lot of electrical energy, significantly affecting the environment.

Coconut palms that grow on the seashore, especially on small islands where there is absolutely no fresh water, produce an abundance of coconuts containing up to a liter of fresh water each. Many stranded sailors have survived on this water on remote islands for long periods of time. Obviously, the recommendation is not to grow coconuts, but to study this wonderful natural process which does not require high temperatures or extreme pressures. The palm tree only needs sunlight and sea water to produce fresh water.

It is not known how this natural process works. It could be based on a type of osmosis. The goal is to reproduce the natural process on an industrial scale at a faster rate.

Mangroves grow in abundance in salt water, using this resource to grow. They have an interesting way of separating out salt, which is visible as clean cubic crystals around the plant's green leaves. This is another biological process that separates salt from sea water at ambient conditions.

There should be no reason why we have to spend large amounts of electrical energy just to separate fresh water from salt water if nature can do it so easily without causing damage to the environment.

XI

NATURAL FERTILIZER

In 2019, the world used 200 million tons of synthetic fertilizer, whose production consumes enormous amounts of fossil energy and thus contributes to global warming through the generation of carbon dioxide. On average, one ton of synthetic fertilizer produces 1.5 tons of carbon dioxide. For this reason, it is essential that we develop an industrial process that does not produce greenhouse gases.

To feed the world's growing population, agriculture increasingly requires chemical fertilizers, which contribute significantly to climate change. The production of fertilizers from nitrogen in the air, converting it into ammonia derivatives, is a chemical process that uses vast amounts of fossil fuels at high pressures and temperatures. It consumes a lot of energy and is accompanied by the inevitable emission of gases, mainly carbon dioxide.

In nature, there are plants that convert nitrogen from the air into their own fertilizer at ambient conditions. In legume plants, such as beans, lentils, peas, etc. and, especially alfalfa, small nodules develop on the roots where bacteria live. These absorb nitrogen from the air and convert it into nutrients such as nitrates, which are derived from ammonia, to stimulate plant growth. For centuries, farmers have known how to reactivate their depleted lands by planting alfalfa and then cutting it (it makes excellent food for livestock). The roots stay in the ground and act as fertilizer for future crops.

This wonderful natural process deserves more in-depth biochemical research in order to learn how these bacteria can activate the nitrogen (N_2) molecule, which is chemically very stable. How can these bacteria form reactive nitrogen derivatives that serve as plant food under ambient temperate conditions? The objective is to duplicate this natural process on an industrial scale, producing natural fertilizer without the emission of the greenhouse gases that are responsible for global warming. The process could involve large tanks, such as fermentation tanks, where these bacteria grow in a solution and air (79% nitrogen) is passed through this liquid.

BIBLIOGRAPHY

It is impossible to present a comprehensive bibliography for this book, because the information is based on

- international news,
- international conferences:
- communications from scientists working directly on the projects involved,
- communications from government agencies in different countries,
- communications from NASA and NOAA (National Oceanic and Atmospheric Administration),
- environmental laws of different countries,
- political announcements, etc.
- and, finally, a basic understanding of physics and chemistry.